TRANSPORT COMMUNICATIONS

TRANSPORT COMMUNICATIONS

Understanding Global Networks
Enabling Transport Services

John Tiffin and Chris Kissling

KOGAN
PAGE

London and Philadelphia

Publisher's note

Every possible effort has been made to ensure that the information contained in this book is accurate at the time of going to press, and the publishers and authors cannot accept responsibility for any errors or omissions, however caused. No responsibility for loss or damage occasioned to any person acting, or refraining from action, as a result of the material in this publication can be accepted by the editor, the publisher or either of the authors.

First published in Great Britain and the United States in 2007 by Kogan Page Limited.

120 Pentonville Road
London N1 9JN
United Kingdom
www.kogan-page.co.uk

525 South 4th Street, #241
Philadelphia PA 19147
USA

ISBN 978 0 7494 5070 0

British Library Cataloguing-in-Publication Data

A CIP record for this book is available from the British Library.

Library of Congress Cataloging-in-Publication Data

Tiffin, John.
    Transport communications : understanding global networks enabling transport services (nets) / John Tiffin and Chris Kissling.
        p. cm.
    ISBN-13: 978-0-7494-5070-0
    ISBN-10: 0-7494-5070-3
    1. Communication and traffic. 2. Transportation. 3. Telecommunication. I. Kissling, Chris, 1940– II. Title.
    HE151.T54 2007
    384'.043--dc22

                                    2007022194

Typeset by JS Typesetting Ltd, Porthcawl, Mid Glamorgan
Printed and bound in Great Britain by MPG Books Ltd, Bodmin, Cornwall

Contents

List of figures   *x*
Foreword   *xiii*
Preface   *xv*
Acknowledgements   *xvii*

**1 Transport plus communications equals globalization:
an overview**   **1**
Introduction   2
Telecommunications   2
Transport   4
Human communications   5
IT   7
Globalization   9
Transport communications   10

**2 We are not trees and we are not sharks: transport
communications theory**   **13**
Introduction   14
We are not trees and we are not sharks   14
Organization theory   16
Systems theory and cybernetics theory   19
Information theory   20
Network theory   23
Fractals   27

Nodes 31
Hubs 32
Location theory 33

3 The networks enabling transport systems (NETS) 37
Introduction 37
Abstract and concrete networks 38
NETS 38
What is transported: people and freight 40
The senders and receivers of transport 41
NETS intelligence: transport services providers 43
NETS 1: transport infrastructures 43
NETS 2: traffic networks 45
NETS 3: regulatory networks 46
NETS 4: communications networks 46
NETS 5: auxiliary services providers (ASPs) 47
NETS 6: skills networks 48
Transport communications theory 49

4 First order meaning: clear transport communications 51
Introduction 51
First order signification 52
Codes and signs 53
Information and redundancy 56
Transport signage 57
Perception 59
Signage heuristics 61
Wayfinding behaviour 62
Signage and IT 63

5 Second order meaning: fantastic transport communications 65
Introduction 65
Second order signification 66
Critical theory 67
Symbolic interactionism 68
Cognitive dissonance 70
The aesthetics of transport communications 74
First- and second-order meaning in organizational
communications 74
IT and second order signification 75

6 **Transport paradigms and the episteme of globalization**   77
   Introduction   77
   Paradigm   79
   Transport paradigms and syntagms   80
   The transport studies paradigm   81
   Epistemes   83
   Paradigm shifts and epistemic shifts   83
   Paradigm shifts in transport at the site level (1810–2010)   85
   Paradigm shifts in transport at the urban level (1810–2010)   85
   Paradigm shifts in transport at the national level (1810–2010) 87
   Paradigm shifts in transport at the global level (1810–2010) 88
   The epistemic shifts of the 1860s and 1960s   88
   The episteme of globalization   92
   Paradigms and the NETS   92

7 **Intelligent transport: the new communications technologies** 95
   Introduction   96
   Nanotechnology   97
   Clever clothes   98
   HyperReality   100
   Artificial intelligence (AI)   103
   New technologies and epistemes   105

8 **Seeking space: water and air transport**   107
   Introduction   108
   The fractal levels of water transport networks   109
   The fractal levels of air transport networks   111
   Infrastructure networks in water and air transport   113
   Traffic networks   116
   Skills networks   119
   Regulatory networks   121
   Communications networks   122
   Auxiliary services providers (ASPs)   124
   Transport services providers (TSPs)   124
   Seeking space   125

9 **Driven: land transport by road and rail**   129
   Introduction   129
   The fractal levels of land transport networks   130
   Land transport infrastructures   131
   Traffic networks   133

Skills networks 135
Regulatory networks 136
Auxiliary services providers (ASPs) 136
Communications networks 137
Driving as communicating 139
The future for land transport 142

10 Walking the walk to talk the talk: pedestrian transport 143
Introduction 143
Walking talking transport communications 144
The fractal level of the site 148
The fractal level of the activity space 148
The fractal level of activity surfaces 149
Ports, portals and protocols 151
Pods and packaging 152
Technological extensions of pedestrian transport 153

**11 Communications for transport logistics and global
supply chains 157**
Introduction 157
Communications within supply chain management 162
The 'before transport' phase 163
The 'in transport' phase 164
The 'after delivery' phase 168
Third party logistics 170
Travel agents as 3PLs 170
Global containerized freight 170
Case study: the cool chain for meat 171
Conclusions 175

12 Troubles in transport 177
Introduction 178
Terrorism 178
War and genocide 179
Threats from the natural environment 181
Danger from disease 183
Automania 185
Cattle class 185
Boat people 190
The perfect storm 190

13 **Scenarios of globalization** **193**

Introduction 193
The doomsday scenario 194
The low carbon economy scenario 195
The u-scenario 195
Smart villages and clever cottages 197
The matrix 199
Seacities and skycities 203
Powered pedestrians 203
The personal pod 204
Which scenario? 205
The global episteme 206

References *209*
Index *217*

List of figures

1.1	Transport and communications overlap	3
1.2	A concept schema of transport communications and globalization	4
1.3	Hierarchy of human communications	7
2.1	An organizational chart following Weber	17
2.2	An organizational chart following Fayol	18
2.3	Shannon's model of a communications system	21
2.4	A model of a communications/transport system according to systems theory, cybernetics and information theory	23
2.5	The island of Sentosa	25
2.6	The transporter terminal at Sentosa	25
2.7	Fractal shift	26
2.8	Node B of the aerial transporter at the Singapore end	26
2.9	Ring networks nested in a ring network	27
2.10	A bus network	28
2.11	A bus network inside a building	28
2.12	An aeroplane arrives at its destination and is linked to an airport	29
2.13	The Sierpinski Triangle	29
2.14	The signpost	30
3.1	The NETS model of transport	40
3.2	An aerial photograph of a remote part of Ethiopia called Janjero	44
3.3	Fractal levels of skills networks	49

4.1	A sign has a signifier, a signified and a referent	54
4.2	An iconic sign	55
4.3	A symbolic sign	56
4.4	An indexical sign	57
4.5	People waiting for their train	58
4.6	Downtown signage in Shinjuko, Tokyo	59
4.7	Distal and proximal stimuli	60
5.1	Relative effectiveness of IT and human communications at the first and second orders of signification	67
5.2	*The Sultana* (1860)	70
5.3	Concorde	71
5.4	The Bugatti Royale	72
5.5	*The Normandie*	72
5.6	Valencia railway station	73
5.7	A modern airport lounge	73
6.1	The relation between paradigm and syntagm	80
6.2	Syntagms of the Spanish railway train paradigm in 2003	81
6.3	A syntagm from the *Shinkansen* railway train paradigm	82
6.4	Paradigm shifts in transport from 1810–2010 at four fractal levels	86
6.5	An idealized overview of paradigm shifts in transport from 1810–2010	89
8.1	The global level network of maritime transport	110
8.2	The site level: water transport in an apple packing factory	111
8.3	A Panama canal lock	114
8.4	Increased use of airspace with GMSC	118
8.5	No frills airline	126
9.1	Kota Kinabalu elevated walkways	134
10.1	An infrastructure with two uses	145
10.2	Charlie Chaplin's tramp	147
10.3	A sewing room	150
10.4	Container shipping	153
10.5	A boat in a Bangkok *klong* or canal	154
10.6	A liner entering Vancouver	155
11.1	The fractal levels of a supply chain from site level to global level	158
11.2	The six NETS	160
11.3	Elements in value-driven chain systems (after Robinson, 2002)	161
11.4	The supply chain matrix	162
12.1	Changi airport automatic check-in system	187

12.2   The interior of a full aeroplane   189
13.1   The old packhorse bridge and 19th-century bridge   198
13.2   The problem with first generation high-rise buildings
       in Singapore   201
13.3   The Twin Towers building in Kuala Lumpur   202
13.4   The matrix city   202

Foreword

Goods, people and information have been likened to solids, liquids and gas respectively. Despite their different composition they have much in common. Similar network structures underpin movements of goods, people and information at all geographical levels from the local to the global. Consideration of all three areas of interest simultaneously allows focus on how they mutually reinforce each other. Further, a range of substitutions can be examined as products, such as music and games, once transported as freight, are now transmitted as information; and some, but not all, passenger journeys to retail centres have been superseded by teleshopping.

Paradoxically, attempts to study goods, passengers and information have focused on differences between them rather than on any commonalities. Consequently, three kinds of specialists have emerged who concentrate on freight logistics, passenger transport and communications. On occasions, freight logistics and passenger transport topics are encompassed at the same conferences, albeit in separate sessions, and sometimes explored within the same transport journals. However, communications is invariably pursued at different venues and has generated a distinctive literature published in a different set of journals. The persistence of this division between the transport and communications literatures and limited attempts at integration of the subject matter makes it difficult for transport and supply chain managers to appreciate the entire picture and anticipate trade-offs within the world in which they have to operate.

In a bid to overcome this impasse, John Tiffin, a communications specialist, and Christopher Kissling, a transport analyst, combined forces to produce this book. The reward for leaving their comfortable academic specialism to explore the interrelationships between transport and communications is to provide deeper insights into the nature and direction of globalization. These insights – embodied in the lucid text, creative diagrams and striking illustrations – emphasize that improvements in transport enhance communications and vice versa. The explorations are invaluable for transport and supply chain managers charged with the task of meeting changing consumer demands by accommodating a veritable swag of new transport improvements and an array of path-breaking information technologies. In guiding their future visions, the opportunity is also taken by the authors to allude to the likely impact of the impending fusion of IT with biotechnology and nanotechnology.

A recurrent theme is the emphasis on the centrality of network analysis highlighting flows, nodes, hubs and land use surfaces. Although cast aside for several decades, network analysis is making an important return to favour as an analytical technique in a world underpinned by a global hub-and-spoke system. Appropriately, in a planet pock-marked by climate change, warfare, terrorism, pandemics and urban gridlock, John Tiffin and Christopher Kissling highlight the vulnerability of the global hub-and-spoke system to natural and human intervention. In particular, major hubs, notably London, New York, Tokyo, Hong Kong and Singapore, play a pivotal role in integrating freight, passenger and information flows and any disruption can have far-reaching consequences on the global supply chain. This study therefore is to be commended as it provides an important global context for transport and supply chain managers exploring the synergies between transport and communications, establishes significant global issues ripe for analysis and offers an appropriate conceptual framework for studying them.

Peter J Rimmer
The Australian National University, Canberra

Preface

This book is for transport managers who need to make decisions about tomorrow's transport. It is about how the relationship between transport and communications leads to globalization.

Transport and communications are normally studied separately. This produces two sets of professionals neither of whom really understand the other. Yet transport cannot happen without communications and the two processes function in similar ways. Their purpose is to move things across space via networks. The difference is that one moves information and the other atoms. The first part of this book shows how transport is made possible by the interaction of transport and communications networks. It provides a three-dimensional matrix called NETS (networks enabling transport systems) that can be used as a guide to improving efficiency in transport management. NETS applies to land, sea and air transport at every level from the international to the national and the urban. It even applies at the mini-levels of buildings, rooms and work surfaces. NETS is, therefore, a tool for reviewing how the different modes and levels of transport can be integrated and automated in supply chains.

Improving transport improves communications and improving communications improves transport. The last half-century has seen enormous improvements in transport efficiency with the shift to diesel engines, containerization, jumbo jets and motorways. This book focuses on the future impacts on transport of the next generation of communications technologies such as broadband internet, wearable computing, artificial intelligence (AI) and HyperReality. It also looks at the impact on both transport and communications of nanotechnology. However, the positive spiral of development that comes from the interaction of transport and communications is in danger of becoming chaotic. The penultimate chapter addresses the issues of global warming, terrorism, pandemics, sustainability and carbon-based fuels. The final chapter looks at transport scenarios for the future which should give anyone associated with planning and designing transport pause for thought.

Acknowledgements

Different iterations of this book were tested on postgraduate students who used it as a set text and told us what they thought of it. We are indebted to them. They came from all over the world and from every mode and level of the transport industry and so gave us a unique global perspective. Many of them were actively engaged in transport management and took the opportunity to apply the ideas in the book to transport proposals and then told us what happened. Their concern to understand the growing conjunction of transport, communication and globalization encouraged us to continue.

We are grateful for the exposure to international thinking on globalization issues that came from meetings of the APEC Transportation Working Group, the Pacific Economic Cooperation Council's Transportation Forum, The Eastern Asia Society for Transportation Studies (EASTS) and The Chartered Institute of Logistics and Transport. We also owe a dept of gratitude to Barry Fairburn and Ross Robinson for their in-depth reading, to Alison Kissling and Christina Sahlin for the artwork, to Robin Wade and Helen Kogan for bringing the book to print and to Maxine Kissling for her advocacy of simple English. We particularly thank Margaret Allan for bringing clarity to the book through her creative editing.

Several chapters include extracts from *Travellers' Tales*, which are just that: travellers' tales. They are from the authors' own experience and like all travellers' tales they are, or should be, perfectly true.

Transport plus communications equals globalization: an overview

As information technology makes it possible to describe and handle increasingly complex transport systems the need to establish a conceptual framework that can serve as a common background for discussion and analysis becomes evident. (Sjosted, 2002)

Ninety years ago our Union's name was the Brotherhood of Railway Clerks. Then in 1919, we became the Brotherhood of Railway and Steamship Clerks, Freight Handlers, Express and Station Employees. The name was expanded more when, in 1967, Convention delegates added the word 'Airline'– making us the Brotherhood of Railway, Airline, Steamship Clerks, Freight Handlers, Express and Station Employees, since then, our Union has welcomed into its ranks the members of... the Transportation-Communications Employees Union (once known as the Order of Railroad Telegraphers), ..., as these other groups merged, strengthening the Union and building it in its diversity, the question of adding their names kept coming up. Many thought that rather than making the name even more unwieldy, we ought to find a way to simplify, to express our unity... Delegates to the 1987 Convention found the solution. Today the Transportation Communications International Union, known as TCU, includes us all. (TCU)

Introduction

On 11 September 2001 two giant jets flew into the twin towers of the World Trade Center in New York and sent them crashing down. Vehicles of global transport became vehicles for global communications to deliver a message that transcended languages and at a stroke changed the world. Transport and communications are symbiotic. They thrive on each other. Improve transport and you improve communications. The improved communications will then further improve transport and so on. This book explains how this works and how it can be used to strengthen management. It shows how, down the ages, humans have used this causal loop to extend their control over territory until today they trade, travel, communicate and struggle for control across the whole of the earth. It is the conjunction of transport and communications that has made globalization possible. But the study of complex systems shows that such cycles of development have the potential to become chaotic (Koen, 2006). One of the great arguments of our time is whether the accelerating growth of transport and communications bears responsibility for climate change and for the conflict of cultures that brought about the events of 11 September. We desperately need to understand the dynamics of the interaction between transport and communications, especially the new breed of communications that comes with the next generation of information technology (IT), if we are to manage transport wisely and well. This chapter lays out a concept schema within which the processes can be studied.

Telecommunications

Telecommunications are ways of communicating across space without physically moving the message or the messenger. They make communications independent of transport and this would seem to

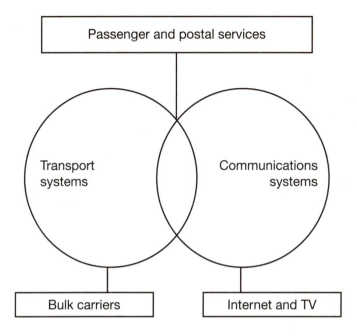

Figure 1.1 Transport and communications overlap

Transport systems that carry raw materials do not normally have a communications purpose, and communications systems that are based on sound waves and the electromagnetic spectrum do not need transport systems, but there is an overlap when communications have atomic form and require transporting.

refute the idea of a fundamental dynamic between transport and communications. Why bother to travel to see someone when you can telephone? Why use transport-based snail-mail when you can e-mail? The symbiosis between telecommunications and transport is subtle. The telephone makes it easier to call a cab. Parcel post has vastly increased since the internet made it so easy to order things by mail. The mobile phone makes people more mobile.

Primitive forms of telecommunications such as drums, smoke signals, rockets, flags, guns and lights have been around since time immemorial. It was in association with steam-powered railways and shipping from 1840 onwards that modern systems of electronic telecommunications developed. Telegraph wires strung alongside railways provided information on railway traffic. The first use of wireless was by the British Navy to maintain contact with its global fleet. The exponential

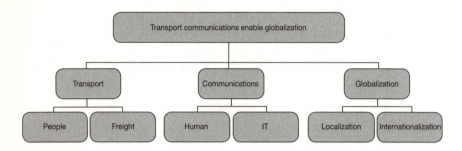

Figure 1.2 A concept schema of transport communications and globalization

Globalization is enabled by the development of transport and communications. Transport systems exist to move people and freight. Communications until recently have primarily been for and between people, but with the advent of IT they can also be for and between machines.

growth of transport systems over the past 150 years stimulated the growth of telecommunications. Now telecommunications are being linked to computers to generate a new surge of growth that will have an enormous impact on transport. Figure 1.2 shows the basic relationships between transport, the different forms of communication and globalization.

Transport

The word derives from the Latin 'portare' to carry and 'trans' across. Transport means to carry something across the space between one place and another. What is moved in transport is matter. This can take the form of manufactured goods, liquids, gases, raw materials, magazines, mail and people. All of these have a fixed physical presence that can be quantified by weighing or measuring volume or by counting the number of units. Transport systems can carry quantifiable amounts of stored information in the form of mail, books and newspapers. They also carry the messages in those letters, books and newspapers that are metaphysical and have no volume or weight.

Transport systems can be classified according to the media they move through and what is transported. There is a basic distinction between land, water and air transport and between people and freight.

It takes energy to transport matter through a medium. When transport depended on people and animals the energy they needed came from the food they ate. When they used sails it came from the wind. The Industrial Revolution meant that the energy that drove transport came primarily from coal and after the 1960s from oil. Coal and oil release gases that pollute the earth's atmosphere and contribute to global warming, but then so too do humans and other things. At issue is not whether the earth is warming. That is measurable and becoming patently obvious. It is whether burning oil for energy is a significant factor in the current phase of global warming such that reducing the use of oil could slow or reverse global warming. A communications war is being waged in the scientific community between climate changers and deniers as they battle for the hearts and minds of the voters in the countries that use the most oil (Pearce, 2006). The outcome will profoundly affect transport.

The term transport is used here to refer to the systematization of transport by humans for purposes determined by humans. This is what distinguishes transport studies from the study of natural acts of transport by rivers, seas and winds that is the province of geomorphology and the explorative acts of travel by individuals that is the province of history. Systematizing transport purposively is only possible if all the components and subsystems are interconnected. The design, development, control, management and governance of transport systems requires communications systems. Until now, transport communications have been primarily conducted by humans, but the conjunction of computing and telecommunications technology is changing this and it becomes necessary to distinguish human communications from that generated by IT.

Human communications

Communications, like transport, is concerned with the movement of things across space with the difference that what is moved in transport is atoms and what is moved in communications is information. Like transport, communications is also concerned with the beginnings and ends of movement where, in communications as in transport, what is moved is stored and processed. The distinguishing factor in human communications is that what is communicated starts and finishes in human heads and the processing and storing that takes place there is called thinking and remembering.

Like all living beings we interact with our environment physically as we push our way through it and subsume parts of it by breathing, drinking and eating. Like all living creatures we also interact with our environment and our fellow creatures by communicating with them. However, unlike any other living creatures, or so we believe, we can also communicate with ourselves about ourselves and how we communicate with the environment and even about how we communicate with ourselves about how we communicate with the environment (which is what you are doing now). We can in fact go off into our heads and communicate with ourselves regardless of our physical environment. We can create and inhabit virtual worlds in our heads and transport ourselves through them and even believe in them. Milton wrote:

'The mind is a place that of itself can make
A heaven of hell, a hell of heaven' (*Paradise Lost*)

What we cannot do yet, however, is explain how communications begin and end in our heads and why we say, write and do the communicative things we do. Neurologists with special equipment can see the flickering networks of neurological discharges that take place in our cortex as we think and communicate and assume that there is a correlation, but as yet they do not know what it is. Our knowledge of human communications is where medicine was 200 years ago. There is no central theory that can explain human communications as satisfactorily as modern medical theory explains disease, though literally hundreds of theories provide explanations of different aspects of them (Heath and Jennings, 2000; Littlejohn and Foss, 2004). Most of them are straightforward common-sense explanations of cause and effect and this book draws on them where they seem to be relevant to transport communications, but there is no unified theoretical explanation of communications that has scientific rigour.

Human communications can be seen as taking place in hierarchical levels (Capella, 1988; Hawkins, Wiemann and Pingree, 1988). Beginning with the intrapersonal communications that go on in the heads of individuals, there are the levels of interpersonal communications that take place between two people in some kind of relationship, the communications that happen in groups, the formal and informal communications of people in organizations, the communications within and between nations and cultures and finally the new level, that of the communications that constitute globalization.

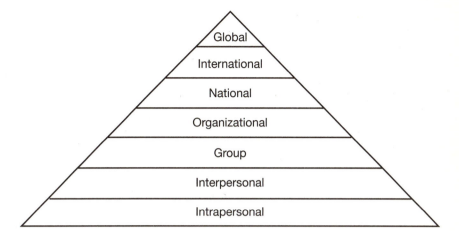

Figure 1.3 Hierarchy of human communications

Early human communications would have been face to face, and so limited by the distance in which the communicants could see and hear each other and the fallibility of human memory. The history of communications technology is one of extending the distance over which people could communicate, the speed at which it could be crossed and the length of time the information communicated could be stored and made accessible. Transport systems are one of the ways of extending the distance over which people can communicate.

IT

The term 'information technology' was first used in 1981 to refer to the growing conjunction of computer and telecommunications technology for storing, transmitting and processing the content of communications as bits of information (Watson and Hill, 1996). The more ambiguous terms 'information communications technology' (ICT) and 'new information technology' (NIT) are also used. The linking of telecommunications to computers makes IT distinctive because it means that there is a technological alternative to human communications. The comprehensive act of transmitting, storing and processing communications that was once the prerogative of humans can now be done independently of humans by machines.

The watcher stood on a knoll where he had a good view of the battle. His horse tethered behind him, his long glass to his eye, he had been there all day as the battle swung first this way then that across the field of Waterloo. Late in the afternoon, the two exhausted armies faced each other in stalemate. Then, in the far distance the watcher saw new movement. The Prussian Army under General Blucher was arriving on the side of the allies. The watcher swung his long glass to the rise where Napoleon and his staff stood and saw Napoleon climb into his carriage and flee the battlefield. The French had lost. Mounting his horse the watcher galloped hard for Ostend where a yacht was waiting. It took the watcher across the Channel in the teeth of a rising gale. In the first part of the following day a relay of horses took him to London. There he made his way to the Stock Exchange. Exhausted, travel-stained and with a sombre expression he walked across to where Nathan Rothschild stood stony-faced. All eyes were riveted upon the pair. No one else in London knew the outcome of the battle. *(Travellers' Tales)*

Had Napoleon won or lost? Yes or no? The answer amounted to one bit of information. Quantitatively, this is the smallest possible unit of information, yet the value of that one bit of information was enormous.

Nathan Rothschild listened to his spy and, without a trace of emotion, made a sign to his agents who immediately began to sell his British shares. The market went mad. Everyone assumed that Napoleon had won and that British shares were worthless. They wanted to sell them at any price. It was not until later in the day, as church bells began to toll a victory, that the stock market realised that Rothschild had been buying back British shares at rock bottom prices. *(Travellers' Tales)*

A 'bit' is an abbreviation of 'binary digit' and there are only two of these: 0 and 1. This system is a way of quantitatively measuring information without regard for the qualitative issues of the meaning it carries. Looking at communications as so many bits of information is like looking at the payload of an aircraft in terms of the number of people it seats as distinct from whether they are tourists or terrorists.

Bits are the basic units of information used by computers to store and process information. Once information has been translated into bits, then communication can be done without human agency. The significance for transport is that in the past all transport control functions, decision making and 'thinking' was done by humans. That is changing. With IT, it becomes possible to have communications systems where information can be transmitted, stored *and processed* independently of human beings. It means that increasingly transport is conducted by IT.

Globalization

The term 'globalization' is widely used to refer to the drive towards free trade at a global level that is taking place under the auspices of the World Trade Organization (WTO) with the encouragement of developed countries (Wolf, 2004). Free trade, however, is beginning to signify more than lifting tariffs so that cargoes can flow freely across national boundaries. The idea of globalization can also be applied to the free flow of people, finance, information and services around the world (Bhagwati, 2004).

The development and improvement of global transport and communications infrastructures make globalization possible. They integrate the world. A glance at the clothes we wear, the implements we use, the food we eat, the jobs we do and the growing variety of people around us shows how far the process of globalization has come. Forecasts for transport and communications developments show no signs of slowing the speed and efficiency in the movement of people, goods and information. What are the limits to globalization?

Unlike transport and communications, globalization has no physical expression. It is a phenomenon that we have come to believe in, and because we believe in it, we institutionalize it. So we now have the WTO, The United Nations (UN), The World Court, The International Telecommunications Union (ITU), the International Civil Aviation Organization (ICAO) the International Maritime Organization (IMO) and the Association of International Automobile Manufacturers (AIAM). These institutions regulate and organize transport activities on a global scale and interact with and support each other. Gradually they are creating a comprehensive organizational structure of concepts, rules and regulations that determine how people interact with each other on a global scale. This does not replace the organizations that integrate transport at regional, national and local levels, but rather adds a new level. And there is a need for adjustment and adaptation between the old levels and the new. Transport services that seek to operate globally find that they need to remember the adage 'think global act local' and design first for the international market (globalization or internationalization) and then adapt to specific local markets (localization).

Transport communications

At the beginning of this chapter are two quotations. The first is from a management perspective calling for a conceptual framework that will link IT with transport. The second is from a union perspective describing how transport and communications workers have over the years come together to form the Transportation Communications International Union (TCU). There is a need at both levels to understand transport communications.

Transport communications is the study of the relationships between transport and communications, in search of a theory that can explain the dynamics involved. Such a theory should make it possible to design, develop, manage and control transport communications more effectively and efficiently. It would also cast light on the socio-cultural and political issues that attach to transport.

Geography texts 50 years ago had chapters on 'Trade and communications' (Briault and Hubbard, 1957) that assumed that the movement of information was the same as the movement of goods and people (Carey, 1989). In pre-television, pre-air travel days, ships and trains carried people and mail as well as freight. In an era of European empires, the flow of trade and communications was seen as a natural movement of people, goods and ideas between the parent country and its empire. Raw materials, whether they were people or goods, flowed one way and finished products the other. This basic pattern continues in the flow of trade and information services between the developed and the developing world and gives rise to the anti-globalization movement (Spoor, 2004).

The objective of many transport systems is to provide the means of communication. A multitude of transport systems are involved in the distribution of books, magazines, newspapers, films, cassettes and discs. People drive or take buses to cinemas, theatres, schools and libraries, all places dedicated to communications. Tourism takes people to experience places through myriad acts of communication. Most work or leisure involves communicating and transport enables it.

Telecommunications does not involve transport. Can we also have transport without communications? Is there any communication involved in the solitary ships that trudge the seas with inert cargos of oil and ore or the long-haul transport of timber by road and rail? Such transport goes its way without seeking attention. Yet it communicates, albeit in a minor key. It may only be a flicker on a radar screen, a routine radio message, or the impact on the imagination of some lonely soul watching a ship go hull down or hearing the wail of a freight train in the night. Even the pariah ships and trains that carry toxic waste and seek the remotest routes to avoid communication activity become world news. Acts of transport are observed and thereby communicate.

Transport systems, because they are systems, cannot function without internal communications systems that link their subsystems and external communications systems that enable them to communicate with their environment. This is axiomatic of all goal-seeking systems. Communications subsystems are always present in transport systems. In the past the communications systems in transport were controlled by people. Cars by themselves were things that could not signal that they needed maintenance or that someone was standing in their way. They had no knowledge of highway codes and could not read the signs on the road. This is changing. Now cars can navigate for their owner and will not move an inch until the driver has fastened his or her seat belt. The growing involvement of IT in every mode and aspect of transport requires us to re-address transport communications.

The interaction between transport and communications is as old as the human species. For at least a million years we have been walking and talking and carrying and then we discovered how to get machines to improve our performance by carrying more, further and faster than we can. Now the machines are acquiring intelligence. The change is radical. It does not just affect the way we drive vehicles. Increasingly what happens at the beginnings and ends of acts of transport is extended by IT into the transit process. Passenger lounges are clogged with business people staying on the job as they travel, by using their mobile phones and the internet. Telemedicine makes it possible for ambulance services to link to teledoctors so that they can be continuously involved in medical situations from the moment they are called to the moment they deliver patients into emergency care.

The continuance and growth of any community depends upon its transport and communications infrastructures. The positive cycle in which good transport systems lead to good communications so that the society they serve survives and prospers gave rise to the empires of Rome, Spain, Britain and the United States. However, it is equally possible to have a cycle in which poor transport systems lead to poor communications systems that lead to even worse transport and the decline of states and the fall of empires, as happened in many places with the fall of the Roman, Spanish and British empires. The dramas of history are set by the transport and communications infrastructures of their time. When these infrastructures change, history changes. Today, we are faced with the prospect of changes in transport and communications on a scale that rivals that of the Industrial Revolution. The internet is in its infancy. Waiting in the wings are augmented reality, HyperReality (HR), teletranslation, robots, artificial intelligence (AI) and nanotechnology. As these technologies mature and converge in transport and communications, the world will shift into a new era with new concepts of community, control and empire.

How do we understand the relationship between transport, communications and the growing orbit within which they take place? The bodies of theory that explain human communications, telecommunications, transport and globalization exist as separate fields of academic enquiry and professional proficiency. They each explain the specific phenomena they address from within their own field of interest. There is no unified explanation that addresses the overall relationship between communications and transport. This book begins to examine the grounds for a uniting theory by looking for what communications and transport have in common. In the case of telecommunications and transport systems this is a rational process because they are both constructs of human logic with a common basis in systems theory which is a human invention. We design and build roads and string telephone wires along them and they work the way we want them to because they conform to the ideas we had for them. In the case of human communications and globalization, however, the matter is not so simple because, although we cause them to happen, we do not necessarily do so rationally. They are as much products of human illogic as logic. They happen and, like the events of 11 September 2001, we do not clearly understand why. Seeking to do so is a long-term aspiration for transport communications.

We are not trees and we are not sharks: transport communications theory

The original model (of transport) was, not surprisingly, an engineering one, arising from technological dominance of rail-born transport in the railway age. Its strength has been extended by the importance of road construction in more recent years and its assumptions are basically those of the producer who can afford to wait for the customer to come to them. During the past 40 years the growing crisis of public transport and increasing concerns of governments with issues of transport policy have led to the development of an economic model which has had relatively little interaction with the pattern of technological constraints and opportunities. This model, while taking account of consumer demand, has found it difficult to allow for seemingly irrational behaviour on the part of human beings unwilling to act as expected by the notion of economic man. Thus a third model, founded in the behavioural sciences, is necessary to complement the other two... As time goes on, we may hope to see the integration of all three models thus producing a unified theory of transport. (Hibbs, 2000)

The web of life consists of networks within networks. (Capra, 1996: 35)

Introduction

Hibbs (above) calls for a model of transport founded in the behavioural sciences which in time would be linked to the engineering and economic models to form a unified theory of transport studies. This chapter seeks to establish the basis for such a unified theory from the common ground in transport and communications theory.

Neither communications studies nor transport studies have as yet, any central unified body of theory of their own. They draw their theoretical roots from other fields from practical necessity. Transport studies is about the organization of transport and communications studies seek to explain how to organize communications. Both draw on organizational theory. Transport studies is about the systematization of transport and communications studies is about the systematization of communications. Both draw on systems theory. The design, planning and control of transport, like the design, planning and control of communications, depends upon feedback as to the extent their systems achieve their objectives. In this both draw on cybernetics theory. Transport and communications both deal with movement and the storage and processing of what is moved, but this is conceptualized differently in the two fields of studies. We also find that information theory, which is central in communications studies, has not been adopted by transport studies, although it would seem to be relevant. By the same token, location theory, which is central to transport studies, has not been used in communications studies although its tenets are relevant.

These theories background the way people in transport and communications think and practise, but what in modern thinking should really draw them together is their common foundation in network theory. This chapter outlines network theory as it applies to transport communications and introduces a number of new ideas from fractal theory and complexity theory.

We are not trees and we are not sharks

An esoteric branch of transport that caters to construction companies wanting instant landscapes specializes in uprooting trees and transporting them to new sites to be replanted. Trees can be moved. There are unfortunate people who are ill and spend their lives permanently attached by wires and tubes to their environment. People can be immobilized. However, unlike the creatures of the vegetable world, we did not evolve to draw the means to live from where we stand. We must either take ourselves to where we can find the means for survival, or we must have them brought to us. We cannot exist without transport. Nor can we exist without communications. We need to communicate to breed, to learn, to interact with each other and the environment. We can shout and wave and use the telephone and so communicate without transport systems, but we cannot transport ourselves or the things we need in any organized, systematic, non-random way without communications, and the whole purpose of transport systems such as postal systems is to communicate. Any theory of transport and communications begins with the fact that they are both processes essential to the survival of humans because they move things across space.

We are not sharks. We do not need to be in constant motion to survive. Nor must we always keep the things we need in motion. Many activities such as reading, writing, eating and sleeping are best done at rest. We need to spend periods of time in one place. Transport may be crucial to survival, but so too are periods of rest and stability. Transport is an intermittent activity for humans. So too is communication. We begin and finish watching a film or holding a conversation. Everyone has to get off the telephone some time.

The very idea of transport, of carrying something or someone from one place to another, presupposes that there is a place of departure and a destination. When we arrive at an airport on one plane only to take off in another, what was destination becomes a new departure, but even the longest voyage comes to an end and we and our goods spend time in one place and have to be in some way stored and kept safe.

The movement of something or someone from one place to another requires energy to propel it through the intervening medium and a period of pause requires some provision for storage and security. This is also true of communication. Energy is needed to propel sound waves from one person to another, to transmit radio signals and to carry the mail. Paper, discs and tapes store messages. Without them communication would be ephemeral.

At this point a physicist would be getting restive. Wind and water are agents of transport that have no discernible beginnings or ends. Like our fellow passengers on planet earth, the trees and the sharks, we are in fact in constant motion all our lives as we hurl around the sun at an average speed of 29.8 kilometres per second. If we microscopically examine the matter we are made of, or that we 'transport', we find that it consists of particles that are in constant motion. Transport studies has, through its roots in engineering, a basis in physical science but, as Hibbs notes in the quotation above, transport studies is becoming a behavioural science in which transport is seen as an activity deliberately initiated and systematized by humans within their world view at a particular period in time.

If there had been schools of transport studies in the universities of medieval Europe prior to Columbus crossing the Atlantic, its professors would not have known to teach about air transport, railways or the Americas. They had no concept of global transport. They would have prioritized piracy in the Mediterranean and argued for the upkeep of the declining system of Roman roads. Theory would have been derived from theology. Whether people and freight arrived safely would have been seen as coming under the control of the almighty and requiring extensive use of prayer.

Something like an early school of transport studies was in fact established in 1416 in the Algarve in Portugal by Prince Henry the Navigator (Russell, 2001). Here studies of navigation pushed at the edges of the medieval mindset and provided the theoretical base for the great voyages of discovery out of Europe that ultimately revealed the global nature of the world. Today rocket technology allows exploration of the solar system and we begin to think of colonizing planets. There may come a time when future students of transport studies will look at today's concept of transport as we look back at the concept of transport held in the Middle Ages. Where we seek to integrate the local and the global they will be designing and planning how to integrate interplanetary with planetary transport. In proposing a theory of transport communications it is salutary to wonder what the theoretical basis for this will be.

Organization theory

Transport studies is about the organized movement of people and freight and organization happens through communications. What distinguishes organizational communications is that they make it possible for organizations to exist independently of the people who come and go in them. Organizations have an existence of their own made possible by the patterns of communicative interaction that people within them are expected to observe.

The person who first recognized this was Max Weber (1948) who developed what is known as bureaucracy theory. This theory has had considerable influence since Weber introduced it in 1909 and its prescriptions remain well entrenched in transport authorities and companies. Bureaucracy theory holds that communications in an organization should be conducted according to formal regulations and standardized procedures in a hierarchy that authorizes who is responsible for what actions. This assumes that people have different tasks within the organization, that there are detailed job descriptions, that people are appointed because of their professional competence in these jobs and that communications at all levels of a hierarchy are conducted rationally in pursuit of the organization's goals.

A good theory can be testable by the argument of fallibility (Popper, 1959). To imagine a transport organization that did not pursue its goals rationally, where staff were not appointed on grounds of their professional competence and so forth, is to appreciate the strength of Weber's theory.

In 1916, French factory owner Henri Fayol first published his work on general industrial management (Fayol, 1949). The principles of management it prescribes, with their concern for analysing goals and tasks, designing strategies and identifying resources, are still the common parlance of modern managers. Fayol's ideas were similar to Weber's with one critical difference. In place of a rigid vertically organized hierarchy, Fayol argued that in some circumstances there could be communication that bridged equivalent levels in an

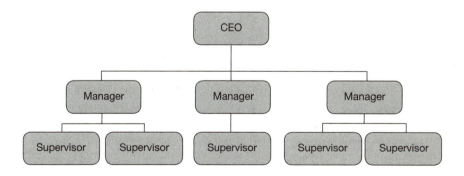

Figure 2.1   An organizational chart following Weber

It is a network in which functions are linked upwards by communication channels.

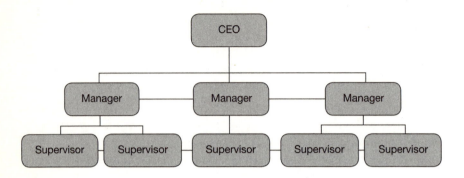

Figure 2.2 An organizational chart following Fayol

While it retains the basic pyramidal shape of a Weberian network, it is more matrixial. Managers and supervisors are linked horizontally. Such hierarchies tend to be flatter in shape than bureaucratic hierarchies.

organizational hierarchy. Today's management would recognize this as flattening the organizational structure.

At about the same time as Fayol was developing his administrative theory in France, Frederick Taylor (1911) was developing the theory of scientific management in the United States. There are similarities with the French and German ideas, but what was special about Taylorism, as it came to be called, was the emphasis it placed on the specialization of labour and the precise measurement of effectiveness by time and motion studies. Taylor provided the rationale that led to Fordism where factory workers do routine repetitive tasks on parts of a product rather than being involved in making the whole product in the way of traditional craftspeople. Until Ford opened his factories, automobiles were individually crafted. With Taylorism and Fordism, the craft knowledge that workers acquired through apprenticeship became the design property of management and the rigorous analysis of the tasks involved in turning a design into a product paved the way for automation.

Weber, Fayol and Taylor provided the foundations of what generically is called organization theory (Kreps, 1990). The concern for goals and emphasis on efficiency provided the management philosophy behind the development of industrial societies in the first part of the 20th century. By the mid-20th century, socialist and communist ideologies that were protective of workers were influencing the social structures in which organizations were embedded. Theories of organization became less mechanistic and more ideological with concern for the aspirations and rights of workers. Organizations were conceptualized as having cultures and ideas of participative management, and team work became popular. However, the downfall of communism and decline of welfarism have seen a reversion to the principles of organizational theory stemming from Weber, Fayol and Taylor. This classic organizational theory with its emphasis on networks, objectives and hierarchies readily links to the systems approach that came to dominate the management view of organization in the second half of the 20th century (Schermerhorn, Hunt and Osborn, 2004; Laudon and Laudon, 2005; McNurlin and Sprague, 2005).

Systems theory and cybernetics theory

Everyone involved in transport studies and communications studies subscribes to systems theory and cybernetics theory. Terms such as 'systems' 'objectives', 'goals', 'inputs', 'outputs' and 'feedback' are common parlance throughout the transport and communications industries.

Ludwig von Bertalanffy (1951) founded systems theory and Norbert Wiener (1948) founded cybernetics theory. The two theories complement each other. Systems theory conceptualizes anything that has a purpose as a system, with outputs that seek to realize its purpose, inputs that provide resources for a system to work and subsystems that interrelate to enable the system to function. Systems are hierarchical, in that systems have subsystems and in turn are within suprasystems. The larger entity within which a system operates is the environment in which it seeks to accomplish its purpose. Transport systems and communications systems are purposive systems and so are seen to conform to systems theory.

All purposive systems also function according to the theory of cybernetics. The term 'cybernetics' has its roots in transport coming from the Greek 'kybernetes' meaning the skill of steering, which is at once a communications and a transport function. A boat is a system designed to meet the objective of taking its passengers and cargo from one harbour to another across a stretch of water. One of its critical subsystems is the person who steers. When this person moves the helm it changes the direction the boat is taking and this then causes the person steering to adjust the helm which then changes the direction of the boat and so it goes on, just as in driving a car or any other vehicle. This circular causality is at the heart of cybernetics. When the person at the helm sees that the boat is veering from its course, he or she is getting negative feedback which calls for a correction to the helm to bring the boat back on course. Goal-seeking systems react to negative feedback by taking corrective action and to positive feedback by continuing to do what they are doing. No transport vehicle would last long without a cybernetics system to steer by and no transport system of whatever size or nature would last without some way of knowing when things were going wrong so that they could be put right.

This circular causality loop that controls a specific purposeful system to ensure it achieves its objectives is known as first order cybernetics. It is the driver in a motor car. At the level of 'is it working correctly or is it not and if not then do this' computers can take over the cybernetic processes. Planes can be guided by automatic pilot. Second order cybernetics assumes an observer who takes a step back from a specific case to study the cybernetics of a class of systems as a whole. How are motor cars achieving their objectives as a means of land transport? The feedback here is very positive. People around the world are buying more and more motor cars and finding more and more infrastructure to use them on by voting for politicians who promise to build more roads and subsidize their use. But more and more cars on more and more roads will ultimately become unsustainable and lead to systemic collapse. In second order cybernetics positive feedback can lead to chaos.

Information theory

A theory is an explanation and a good theory is an explanation that is strong enough and clear enough to serve as a basis for designing and planning things according to the theory. In the 1950s, systems theory and cybernetics theory came together with information theory to provide an explanation of how communications systems worked. Although this did not fully explain how human communications worked, it did provide the rationale for the development of computers and for linking computers to telecommunications to create IT. Things designed by applying these three theories, such as the internet and traffic control systems, work with great consistency. However, unlike systems and cybernetics theory, information theory has not been widely adopted as such in transport studies. This is in spite of the fact that information theory is implicit in the study of transport because: (a) information theory is an explanation of movement, (b) transport can only function systematically with information and (c) transport studies does use network theory which is a logical extension of information theory.

Information theory has its origins in the work of Claude Shannon (1916–2001). He illustrated the path for transmitting information over space with a diagram which has come to dominate the way people see communication as point-to-point movement across space with one point having the function of source and the other of destination (Shannon and Weaver, 1949). This, of course, is exactly how transport is seen in transport studies.

Shannon created his model while he was working in the research laboratories of the Bell Telephone Company. He was concerned with the basic problem of how to transmit information from one place to another place efficiently so that it arrived as nearly as possible in the same condition as it was despatched. Any transport engineer could identify with such an objective and with Shannon's basic model of a source, a destination and a channel between them. Transportation networks are based on the fundamental dyad of an origin and a destination linked by a route (Kansky, 1963). The differences lie in what is moved and the terminology used. Where Shannon was thinking of moving bits

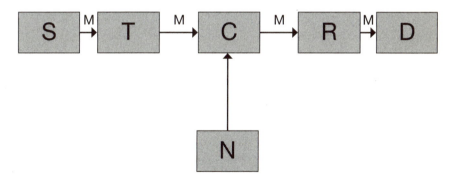

Figure 2.3 Shannon's model of a communications system (Shannon and Weaver, 1949)

An information source (S) originates a message (M). A transmitter (T) encodes or embodies the message in the medium (C) used to channel the message to a receiver (R) that decodes and makes the message available to a destination (D). (N) is the source of noise that is added to a message in transmission.

of information, transport people think in terms of moving matter and where communicators talk of transmitting and receiving, transporters say they are despatching and docking. Shannon's concept of 'noise' (Shannon and Weaver, 1949) also sounds odd in the context of transport. It refers to what happens to information during transmission. A similar principle exists in transport: it is impossible to move something from one place to another so that it arrives exactly as it left. There is always some change in what is transported or communicated. A difference, however, is that in communications studies what happens in the transmission of information is essentially seen as causing deterioration in a message, whereas what happens in transporting people and goods is essentially seen as adding value. Obviously, change takes place in what is moved as a result of the passage of time, the difference in space and the transport/transmission process itself. This can be good or bad, regardless of whether it is information or atoms that are being moved. A cargo of fruit can be damaged in transport or ripened and messages can be conveyed too soon or too late.

All purposeful transport and communications is done so that what is moved can in some way be processed. Harold Lasswell came up with a modified version of Shannon's model that allowed for this in communications. Known as Lasswell's dictum, it takes the form of a series of questions, the most critical of which is the last:

'Who?'
'Says what?'
'To whom?'
'In what channel?'
'*With what effect?*' (Lasswell, 1948)

Lasswell assumes that all communications has an effect and wants to know what it is, because it is this that measures the effectiveness of communications and relates information theory to systems theory with its concern for objectives and to cybernetics theory with its call for feedback. Lasswell's dictum can be applied to transport. All that is needed is to change 'says' to 'sends'. It would seem that the main problem in transport studies adopting the principles of information theory is semantic.

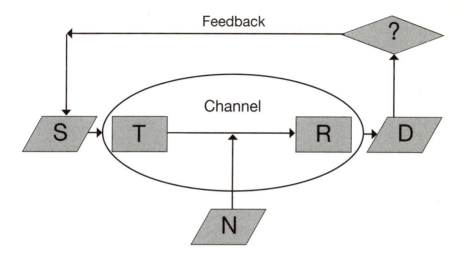

Figure 2.4 A model of a communications/transport system according to systems theory, cybernetics and information theory

The oval encompasses a system that could be a telephone company or a shipping company. S is a customer who wants to send a message or a container to D, a destination. In communications, T and R are telephones and in transport, ports (both are simultaneously transmitting and receiving systems). The channels are telephone wires and cables or shipping routes and a possible source of noise could be a storm that caused a lightning hit on a cable or moved a container. The question mark is Lasswell's 'With what effect?' Did the message or the container get through? A feedback loop carries the answer back to the customer.

Network theory

Here we are back to a theory that both transport and communications studies unambiguously draw on. The idea of a network has been widely applied in many fields and disciplines. It is used in neuroscience to describe the neural networks of the brain (McCulloch, 1988), in epidemiology to trace the links by which diseases are transmitted (Ancel *et al*, 2003), in sociology to describe the patterns of communication between people (Moreno, 1934; Rogers and Kincaid, 1981) and in computer science to describe what goes on inside and between computers. Barabasi (2002) sees networks in the organizing structure of Al Quaida, the stockmarket and cocktail parties. Transport systems are networks and communications systems are networks and since all transport systems are communications systems all transport networks are communications networks.

At the close of the last century network researchers in different disciplines found themselves linking at a global level on the internet. They were in a virtual laboratory where a network of remarkable complexity was evolving exponentially. This shared experience has renewed interest in networks (Buchanan, 2002) and provoked the idea that they may describe fundamental laws of nature (Capra, 1996; Barabasi, 2002; Taylor, 2003). Transport and communications could be described as fundamental processes that manifest and function as networks.

Networks are sets of interlinked nodes. A node is a point where two or more links come together and a link is what joins any two nodes. Shannon's model of communications describes the dyadic relation between two linked nodes. So network theory links to information theory. It also links to systems theory and cybernetics theory and through them, more tenuously, to organizational theory. With network theory a degree of unity begins to emerge.

The Shannon dyad is the basic unit of all networks and the simplest form a network can take. When another link to another node is added, then a network acquires new properties and, as more links and nodes are added, further properties emerge. The extent to which interconnection topologies conform to bus, ring and star patterns affects how networks function, as does the way networks may be linked to networks at different hierarchical levels. Here we come to a phenomenon that was first noted in communications networks (Tiffin and Rajasingham, 1995) but which would also seem to apply to transport networks: fractal shifts can take place at nodes so that nodes can become networks and networks nodes. A city that is shown as a dot on one map can on another map at another scale be shown as a network.

Figure 2.5 The island of Sentosa

Sentosa is a small island (A) off the main island (B) of Singapore. The islands are linked by an aerial transporter. This forms a dyadic transport network in which the islands are nodes.

Figure 2.6 The transporter terminal at Sentosa

Node A is the transporter terminal at Sentosa, where the gondolas are turned around as people get on and off. From the point of view of passengers this is the end of a journey to Sentosa or the beginning of a journey to Singapore.

Figure 2.7 Fractal shift

These people have disembarked from the transporter and are still in the terminal node of the transporter network as they orient themselves. Above them is a noticeboard that points to the paths they can take. From being their destination, node A now becomes for them the central node in a star network of footpaths that lead in every direction across the island.

Figure 2.8 Node B of the aerial transporter at the Singapore end

An elevator links node B of the aerial transporter to ground level (C) where there is a bus station and taxi stand that opens up the road network of Singapore to the traveller. The elevator itself is a simple dyadic link.

Fractals

Fractal geometry was developed by Mandelbrot (1982) to describe natural features with irregular forms such as clouds or trees or mountains in which a pattern can be discerned that keeps repeating itself at different levels. He also applied fractal theory to financial markets (Mandelbrot, 2006) and regular geometrical forms that repeat their shape at different levels. These may be, as in the case of the Sierpinski Triangle, infinitely recursive, or, as in the case of the roundabout shown in Figure 2.9, they may have a finite number of fractal levels (in this case two).

Where fractals fit in transport and communications networks is that a node in a network at one level can become a network in itself at another level and conversely a network can become a node in another network. At the national level of interurban transport networks, cities are nodes. At the urban level of road transport, buildings are nodes. Enter a building and it becomes a network of corridors linking rooms that are then the nodes.

Figure 2.9 Ring networks nested in a ring network

The 'Magic Roundabout' at Swindon (UK) where each of the five nodes in the main ring is itself a roundabout with four nodes.

Figure 2.10 A bus network

Here nodes are linked in sequential order like the stops along a bus route. In this case the bus network is the house nodes linked in sequence by the road.

Figure 2.11 A bus network inside a building

Rooms are nodes linked along a corridor.

Figure 2.12 An aeroplane arrives at its destination and is linked to an airport.

From the perspective of the airline it has made a dyadic link between two airports which are nodes in a network of flight routes. From the point of view of the airport authorities, the airport itself is a network that links ground transport networks to air transport networks and one flight to another. Nested in the airport terminal network are smaller networks such as the air bridge in the photograph, which gets people from the aeroplane to the terminal building and the motorized transport, which does the same for their luggage.

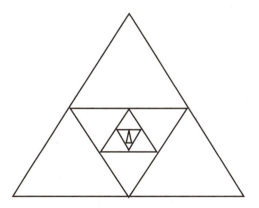

Figure 2.13 The Sierpinski Triangle

This is an infinitely recursive fractal.

Figure 2.14 The signpost

The signpost labels a point where many roads and paths come together. It is a node in many networks. The two arrows at the top point to nodes in the urban road network that links the buildings in a small town in the UK called Pateley Bridge, where the signpost is located. The two signs below point to the next towns of similar size on a B road network. Lofthouse and Middlesmoor are small villages on a minor road network. The bottom sign points to a node in a network of rural paths. People pause at this signpost to work out where they are going and to shift from one network to another. From the point of view of travellers looking at all the possible places they could go to from here, the signpost is the central node in a star network.

Nodes

Nodes exist notionally where links in a network meet. The fractal shift that comes with the way we perceive them can turn them into a network in their own right. Other functions that may be found in transport and communications nodes are as follows.

■ Source and destination. The transport of humans, like communications, is episodic. The motel that is a destination for one day's driving becomes the starting point of the next day's drive. The end of one chapter is the beginning of the next.

■ Gate or portal. A node can exist in two different kinds of networks serving as an entry/exit between them. It can be a destination in one network and a starting point in another. The gate that is an aeroplane's destination is the beginning of a passenger's march through the network of an airport to the taxi rank, which serves as a gate from the airport network where they enter a taxi to start a journey through an urban road network to the destination of their hotel. Here a reception desk is a gate that gives access to the internal network of the hotel. Similarly a letter travels through a hierarchy of networks that are reflected in the address of the recipient which shows their country, their city, their building and finally the person themselves.

■ Switching the direction of traffic. This is when an airport serves as a transit point in a journey. In the early days of telephones, operators would manually switch telephone calls at the local level to the national level and then the international level.

■ Storing or resting what is transported or communicated. This is the function of waiting rooms, warehouses and hotels. It is allowing for the intermittency that exists in transport. The equivalent in communications would be a library or newsagent.

■ Acquiring the energy needed to cross the space between nodes. Petrol stations are critical nodes in a road transport system. Restaurants and bars are critical nodes for the people talking at a conference.

■ Achieving the purpose of a transport or communications system. Manufactured goods are transported to where they can be sold.

People are transported in order that they can work or shop or go to school. Airlines do not simply sell 'bums on seats'. They advertise tourist packages that include accommodation, access to beaches, sports and entertainment as well as a return trip. They try to ensure that the customer has the holiday that was the purpose of the flight.

■ Function of transporting. It sounds contradictory, but nodes are not necessarily fixed facilities such as a bus stop or an airport. They can be floating. This is the case with vehicles such as buses and planes that transport things. People are floating nodes in pedestrian traffic networks and in the communications networks that form on social occasions such as a party.

Hubs

Barabasi (2002) sees complex networks as having three basic topographies. There is the random network where all nodes have approximately the same status. An example would be establishing a circular bus service network where each bus stop is a node linked to the stop before and after it to provide the same degree of access to buses. As networks grow, some nodes may attract more traffic and more links than other nodes. The longer a node exists and the more popular it is, the more links it attracts. Nodes with lots of links get even more links because they already have a lot of links. Barabasi identifies a class of networks he calls 'scale-free' in which most nodes have very few links, while a minority of nodes, which he calls hubs, have a large number of links and there is a graduation of intermediate-sized nodes between these extremes. Barabasi uses the example of the US air traffic system, where a small number of major airports are responsible for most flights, in contrast to a large number of small airports with very few flights and a distribution of medium-sized airports in between.

According to Barabasi, giant hubs make such scale-free networks highly reactive. He refers to the way people with many connections to other people can be regarded as hubs and how this explains the small-world phenomenon in communications: that anybody, anywhere can be linked to anybody, anywhere via six communication links (Milgram, 1967). The idea is critical to the push to eliminate intermediaries in supply chain management.

Barabasi's third kind of network is when a single hub is directly linked to all the nodes in a network in a superhub topography. A television network where one station broadcasts one way to all the television sets within range of its broadcasts is an example from communications. Napoleon, when he made Paris the centre of a giant road system to improve his control of Europe, sought to make such a network, as did Bismarck when he made Berlin the hub of a national railway system that could rush troops to the borders of Germany. Scale-free networks are between these extremes and tend to develop from being random networks to becoming superhubs. Barabasi suggests that scale-free networks may be typical of complex networks such as the internet, and the human genome. However, transport systems can also develop network topographies that conform to the unit norm or bell curve where there are very few extremely small and extremely large nodes and the majority of nodes are of average size. For example, railways seeking to rationalize may reduce the number of small stations so that the majority of stations are intermediate in size. Moreover, transport and communications networks seldom, if ever, function as single unified networks (see Chapters 8, 9 and 10).

Location theory

A summary of the theoretical roots of transport studies cannot conclude without mention of location theory (Nickel and Puerto, 2005). This is an important body of theory in transport from German economic geography that seeks to explain why economic activities are located where they are, and obviously this profoundly affects transport. It has, however, had surprisingly little impact on communications studies despite the fact that many communication activities are economic activities that depend on location, for example the film industries in Hollywood and Bollywood.

Johann Heinrich von Thünen (1783–1850) founded location theory as he sought to explain the factors that determined what kinds of agriculture were best suited to what regions. Alfred Weber (1868–1958) did something similar for industry. The theory provided a rationale for building roads and railways in accordance with economic planning.

Location theory is based on the assumption that transport networks are fixed by economic factors of distance. It has proved reasonably predictive, especially as it has been refined over time to take more and more factors into account (Nickel and Puerto, 2005). However, new location factors have come into existence with globalization and the growing application of IT that combine to cast fixed facility nodes adrift. Think of how the fixed facility node of an office with its telephone and desktop PC has become a floating node with the advent of mobile telephony and laptop computers and how this impacts on transport communications.

A horticulturalist working 100 years ago might have established a cherry orchard at a distance from the nearest market town that was optimal for orchardists according to location theory. This relatively stable situation would encourage the development of suitable transport systems between the orchard zone and the town. Today, the town might have an airport with flights to Japan where good quality cherries are at a premium at certain times of the year. The cost of flying cherries to Japan can at such times be small, relative to their retail price, and the return from doing this may be better than could be obtained from the local market. Modern telecommunications puts Japanese cherry buyers directly in touch with cherry orchards around the world and jet cargo flights can make global transport by air as quick as national transport by road and rail. This also means that the supermarkets in the town traditionally served by the orchardists are, in their turn, no longer dependent for their cherries on local suppliers.

Japanese cherry buyers and supermarket chains survey the world for suppliers and monitor growth in different countries to get the cherries they want when they want them. Cherry growers are exposed to volatile global market conditions. They may have to uproot their cherry trees and replace them with other forms of land use. Cherry trees can even be grown in special containers that allow them to be transported to new sites. Cherry orchards are not exactly floating nodes, but like many of the things that are transported or communicated, they are not as fixed as they used to be.

Heraclitus is reputed to have said that it is not possible to step into the same river twice. This is now the essence of location theory. Every point in space has at any given time a unique value. The story in Chapter 1 of the watcher at the battle of Waterloo illustrates this. The information he had as to who had won had no value on the battlefield where everyone knew the outcome, but it had enormous value on the London Stock Exchange when no one else knew.

3

The networks enabling transport systems (NETS)

We live in a small world where everything is connected to everything else. (Barabasi, 2002: 7)

Introduction

The last chapter sought to summarize the theoretical foundations that transport and communications studies share and to unify them in network theory. In this chapter we identify six kinds of networks as being essential to any transport system. They are: infrastructure networks, traffic networks, regulatory networks, communications networks, auxiliary services networks and skills networks. Any transport system in any mode at any level has to be supported by all six kinds of networks. From these six different kinds of networks we develop a tool which we refer to collectively as NETS (see Figure 3.1). NETS visualizes supply chain activities as a kind of Rubik's cube that can be manipulated to find the optimum path for transport. The tool has, therefore, the potential to become a means for designing, planning, managing and rapidly revising supply chains.

That transport depends on the interaction of concrete networks of infrastructure and traffic is established (Morlock, 1967), as is the need for networks of auxiliary services such as the police, ambulances and fuel stations. What is different about NETS is the recognition of the importance of abstract systems of a communicative nature such as the networks of regulations, skills and communications.

Abstract and concrete networks

Networks can be abstract or concrete and humans have the perceptual ability, as it were, to morph between the two. A traveller looks down at a crossroads marked on the abstract network of a map and then up at the signpost that establishes it is the same crossroads in a concrete network of roads. In transport and communications, abstract networks are regularly transformed into physically concrete networks and vice versa. Maps, plans, timetables and organograms are representations of abstract networks. They describe what is supposed or proposed to exist. People have in their heads an abstract vision of the part of a road or rail network they are using.

The bit of information the watcher at Waterloo transported to the London Stock Exchange had no physical existence in itself. It was a key part of a network of knowledge about stocks and shares in which whether Napoleon won or lost the battle was linked by 'if–then' logic to a fall or rise in the value of British shares. Bodies of knowledge can be conceptualized as abstract networks, in which concepts, facts, assumptions and ideas are the nodes and the links between them are formed by some kind of logical dependency.

NETS

Morlock noted that 'A transportation network is in actuality two networks – one representing the network of fixed facilities and the other the network of vehicle movements on those facilities. There is no necessary one-to-one correspondence between these, so they cannot be treated as one. The major interaction is the restrictions which the fixed network places upon vehicle flows especially with respect to between what places and via what routes vehicle flow occurs' (Morlock, 1967). By the same premise, there are other networks, actual and abstract, that are critical if transport is to take place. Systematic transport, in whatever mode and at whatever level, is made possible not only by networks of vehicle movements and fixed facilities, but also by a further four kinds of network. These six different kinds of network have no comprehensive correspondence with each other and do not form a unified system, but without all these different kinds of network systematic transport could not take place. These networks are as follows.

- Infrastructure networks: networks of the fixed facilities that constitute a physical infrastructure for transport (ports, stations, roads and railways).
- Traffic networks: networks of vehicles that constitute the floating nodes the movements of which in relation to each other constitute the dynamic network we call traffic.
- Regulatory networks: networks of the rules, regulations and schedules that specify how traffic should use an infrastructure and relate to other traffic. Such networks include any plans for any new transport system that specify how traffic will operate.
- Communications networks: networks that inform transporters as to where places are, what regulations are in force, how they should interact with the infrastructure and what dangers or special conditions apply. No purposive system can function without communications systems that allow the flow of cybernetic and control information between and within its subsystems and with its environment.
- Auxiliary services networks: networks of the auxiliary services needed to ensure and maintain the efficient and correct use of the other basic networks (eg petrol stations; police, ambulance, lifeboat and meteorological services).
- Skills networks: networks of the skills needed to operate vehicles in traffic on the infrastructure according to the regulations and to understand and act on transport communications.

These are different kinds of networks. In any one category there may be several similar networks. For example, the auxiliary services networks needed for road transport include networks of petrol stations and police and ambulance services, the infrastructure network of roads may include networks of pavements, tramlines and bus lanes.

These NETS exist in their own right to make transport possible, but for transport to happen there must be someone or something to transport, someone or something that decides to transport it and

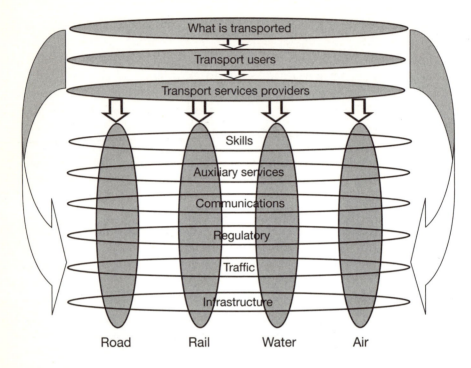

Figure 3.1 The NETS model of transport

Without the six different kinds of network that enable transport there can be no systematic transport in any of the four basic modes of transport. For transport to happen there must be something to transport (people and freight), a transporter (transport service provider) and something or someone to initiate transport.

someone or something that does the transporting. The reference to 'something' deciding and transporting is because increasingly IT is involved in these functions. What follows is a description of how these functions interact with the NETS in a comprehensive model of transport.

What is transported: people and freight

The *raison d'être* of transport are the atoms that someone or some organization wants moved from one place to another. What is transported can have its own intrinsic motivation to be transported because it would in some way be better off or more valuable somewhere else. Travellers have their reasons for travelling and coal in the ground does not have the same value as coal in a furnace.

What is transported may be regarded as an autonomous unit, as in the case of an individual traveller with their own vehicle or a letter from one person to another. However, what is transported can also exist as a network, as when passengers are a tourist party, a family or a football team and want to be together. By the same token containers can be networked together on a ship or dockside according to their destinations and priorities. There is a network in the way teams from different countries are brought together in time for some international event and a similar network in the way tyres from one country, chassis from another, bodywork from a third come together for just-in-time assembly as a car in yet another country.

The senders and receivers of transport

In a traditional railway station the station manager would check that everyone and everything was safely aboard a train and the doors shut and then blow a whistle to alert the driver and wave a green flag to send the train on its way. Down the line at the next station another station manager would be waiting to safely receive the train, its passengers and any freight. These basic functions of sending and receiving what is transported are always there in some form at some level. The station masters dispatch and receive trains, ticket collectors dispatch and receive passengers, passengers dispatch and receive themselves when they board and get off a train, and railway companies inaugurate and close services.

Fords, when the water is up, are good places to watch people deciding whether to initiate transport. They want to cross, but have to decide whether it is too dangerous. Every transport event, whatever the transport mode, whatever the level of transport, be it crossing a stream or crossing the Pacific, involves a management decision to do it that takes into account the risks, the costs and the capability of the different NETS.

Traditionally it is people who make these decisions and have the responsibility for initiating and putting closure on transport, but increasingly this function is being automated. People buy a ticket from a machine, use it to go through automatic turnstiles and get on a train when the doors automatically open. Where decisions to initiate transport are complex and consequential, as in launching a new freight service or a missile attack, decision support systems are used. Unlike the human brain, they improve with every generation.

> The itinerant fruit picker picked up a basket and moved off to the orchard. It was the turn of the traveller. 'Name?' said the old farmer without looking up. The traveller waited until he did. 'Oh, it's you.' The two grinned at each other. 'Get on then, you know the routine.' And the traveller did and he loved it. Up the ladder on a crisp sunny autumn day with his head among the apples. A goodly part of the crop were in perfect condition for picking. Not yet fully ripe but they were going into cold storage where they would slowly ripen on their passage to the other side of the world. He liked to think about the plates they would finish up on and selected carefully, but a voice below interrupted his musings. 'Get on with it! I've got the trucks coming at two o'clock and they will only make one trip. There's a ship due in tomorrow and God knows when the next one will be in.' The traveller nodded and worked faster and less carefully, but the way of working did not suit him and the day was turning seriously hot, so he decided to move on. Later that afternoon he caught a lift with one of the trucks as it left for the packers. He was telling the trucky about the pressure to pick as they pulled up outside the fruit-packing plant. The trucky grinned and pointed to the huge queue of trucks they had joined. They were all waiting to disgorge apples. 'More haste less speed. Looks as though every farmer in the country has the same idea. By the time we get these apples in the store they won't be ripened, they'll be baked.' (*Travellers' Tales*)

The owner of the orchard in the story had made a decision to send a load of apples to the local packing centre based on information about the weather, the availability of pickers and the condition of the crop. But similar decisions were being made by neighbouring apple growers who used the same fruit packer. Then there were the decisions made by trucking companies and independent truckers and by the fruit-packing plant as to the numbers of trucks they can handle over a given period. There are also decisions by shipping companies and their skippers as to when a ship is available and by harbour authorities as to the availability of docking facilities. The scope for the kind of calamity that occurs in the story is high when it depends on traditional human decision making. However, scheduling is increasingly supported by IT-based intelligent integrated decision-making systems.

At the micro level the traveller, under pressure from the farmer's decision to have his crop on the road by the afternoon, is faced with decisions as to which apples to pick and so directly initiate transport for an individual apple, which is finally selected from a fruit bowl by the person at the end of its journey.

NETS intelligence: transport services providers

The intelligence to operate within NETS is in the people who have the skills to drive a vehicle through traffic along an infrastructure from a source to a destination according to the regulations that apply. They apply intelligence to the NETS to make transport happen. They are the drivers and the pilots. In urban transport they are the pedestrians navigating their way through crowded streets.

Traditionally it has been individual people who do the transporting. However, the transport services providers who employ drivers and pilots assume responsibility for transport and in this sense can be thought of on a broader scale providing the intelligence to integrate the NETS. In doing so, today's management seeks quality control and specifies exactly what is expected of their employees. As they do so, they apply neo-Taylorist analysis to the actions that constitute intelligence in transport. What is it that actively integrates the NETS to make transport happen? To the extent that it is possible to answer this question it is possible to automate transport and introduce AI.

NETS 1: transport infrastructures

Figure 3.2 illustrates how a natural network develops from people taking the path of minimum resistance between places they want to go to in pursuit of trade and to communicate. Having developed paths from frequent use they tend to stay on them and improve them. Many road systems evolve in this way. Traffic created a network infrastructure and then the infrastructure channelled traffic. The relationship is like that between a river and its riverbed. They each shape the other. Over time, the network of paths and their use becomes institutionalized. Roads are metalled, cuttings, embankments and bridges allow wheeled vehicles, traffic is controlled, regulations and laws determine how

Figure 3.2 An aerial photograph of a remote part of Ethiopia called Janjero

A path used by people walking or leading mules and donkeys can be seen as a white line running along a ridge that runs from top left to bottom right. The land falls away steeply on both sides of the path, which is the main means of transport and communications for the people who live along the ridge. Another path joins it (top left) which is a link with lowlands for trade. Where the two routes meet is a node in this simple network. To zoom into the node would be to find that it is a hamlet with its own network of paths between the huts. This is a natural transport network formed by the way the landscape channels the needs of people to communicate and trade. The black line running from top to bottom and crossing the ridge route is a boundary line. The white spot comes from the bare ground where people have to wait to pass through the boundary. It is where the use of the highway is controlled.

the network should be used. The network infrastructure becomes an artificial system, designed, built and maintained in response to the needs of the traffic, while at the same time shaping the nature of that traffic.

Infra means 'underneath'. An infrastructure is the underlying structure that supports an activity. The term describes the fixed networks that support societies such as roads, sewers, water pipes, electrical grids and telephone cables. These infrastructures make possible the movement, storage and processing of the things that are essential to civilized life. In a railway system, the infrastructure network consists of the fixed facilities of railway track, bridges, cuttings, embankments and stations. It does not comprise the trains and the people who ride on them. Infrastructures seem hardwired, cast in concrete or carved into the landscape. Morlock (1967) described an infrastructure as a network of fixed facilities. Roads, railways, towns, stations, airports and seaports do not move relative to the traffic that uses them. Or so it seems to human perception. Infrastructures are in fact dynamic from a historical perspective in that they grow or decay and shift their position in response to the pressures from the people that use them. Ports shift downstream towards deeper water as ships increase in size. Airports grow and have to change the positions they occupy. Segments of railways and roads are constantly being repositioned. Some infrastructures actually do have visibly moving links. An escalator is an infrastructure that moves between one level and another while the traffic that uses it may be stationary. In the future, it could be the road that moves while the traffic is stationary on the road.

Transport infrastructures are managed so that cybernetic flows of information about the state of the network, the traffic that uses it and the extent to which it is achieving its purposes can be translated into a corresponding flow of controlling information from management that results in improvements to the infrastructure.

NETS 2: traffic networks

The nodes in traffic networks are the vehicles using a particular infrastructure network. They share the constraints and communications from signage systems that go with the joint use of a stretch of infrastructure. Where infrastructure network nodes are normally fixed, traffic network nodes are normally floating. Vehicles as traffic nodes may alternate between being floating and fixed in that they may stop moving along an infrastructure link and 'dock' with an infrastructure node as when a train stops at a station or a bus at a bus stop.

Traffic networks are dynamic. Vehicles join, leave and change position relative to each other and to the infrastructure. Links intensify with proximity. Ships, aeroplanes and cars that are a long way apart in a traffic network have only attenuated links. As they close with each other, their options for placement in a traffic network are reduced and communications become increasingly critical.

People as pedestrians are nodes in pavement traffic, but when they step into a vehicle, whether as driver or passenger, it is the vehicle that is the node. Traffic networks are traditionally linked by human intelligence and human perception in the vehicle nodes. Interaction between vehicles is by line-of-sight communications. Vehicles see and hear each other and signal and act accordingly. One of the key issues raised in this book is the way the intelligence in traffic could become a function of AI in IT systems which have the ability to 'see' and control all the vehicles in a network.

NETS 3: regulatory networks

Infrastructures are approved, built and operated in accordance with abstract networks of laws, regulations, cost structures and schedules. The vehicles that compose the traffic that uses the infrastructure are manufactured, licensed and operated in accordance with these networks. The networks prescribe at local, national and international levels how transport should be conducted. Without them there would be chaos. They exist as networks in that the nodes are fixed laws, regulations, costs and times that are linked together by logic and antecedence. However, these nodes do have concrete manifestations that link them with the other NETS in such forms as signs that state speed limits, tickets that verify that the right to use a transport system has been purchased and the behaviour of transporters that shows compliance with regulations.

NETS 4: communications networks

There are many kinds of communications networks in transport, though not all are critical in the sense that without them there can be no transport. Modern aeroplanes could not function without radio links to traffic control systems, but they could manage without cabin speaker systems or showing films. What are critical are the basic communications systems within and between the NETS that enable them to function and the signage systems that advise, warn, guide and mandate what people should do in transport, where places are and what services are available. Transport services providers depend on the mass media to advertise their services. There would be no transport if people did not know what was available.

There are two kinds of communication. There is a first order of clear communications and a second order of communications that appeals to the imagination. How this works is the subject of the next two chapters. The kind of communications that are critical to transport are of the first order. Through the mass media people can express their feelings about transport, the lack or excess of its services, how well or badly the laws that govern it are working, how safe or dangerous traffic is and what needs to be done about the skills or habits of those who drive. They also do this in their everyday face-to-face communications, for example as they get on buses or aeroplanes and voice their pleasure or displeasure. This is the communication of feedback and it is essential to any purposive system.

NETS 5: auxiliary services providers (ASPs)

Auxiliary services providers make transport possible, as distinct from transport services providers (TSPs) who do transport. ASPs include gas stations, motels and places to eat and rest. TSPs include bus services, taxi services, couriers and hauliers.

> Travelling fast along a lonely country road the traveller noticed another car approaching and slowed down. The other car gave him a friendly flash of his headlights which he understood to mean that police were around. However, since there were no other cars in sight on the stretch of road, the traveller speeded up again. The approaching car suddenly switched on a blue light and a siren. It was the police.
>
> Standing at the side of the police car while the speeding ticket was made out, the traveller began to argue over the speed he had been clocked at. A third voice entered the discussion over the police radio to say that he had been clocked travelling at high speed long before he met the police car. How on earth could they know? The traffic policeman grinned at him and pointed into the sky where a spotter plane was circling. (*Travellers' Tales*)

The cars of the policeman and the traveller in the story were nodes in the traffic network, but the traffic policeman was also a node in an ASP. Without police, fire and ambulance services, road networks would be dangerous. In this case the police used planes as well as cars. Its traffic network crossed over two modes of transport and was linked by radio.

NETS 6: skills networks

In a skills network the nodes are specific skills that are linked to each other by logical dependency. In order to be able to do a particular skill a person needs to have certain other skills. Robert Gagne first showed how skills could be networked to each other in a 'learning hierarchy' which established the sequence of subsidiary skills a learner needed to acquire to move from what they already knew to what they wanted to know (Gagne, 1965). Tiffin and Rajasingham (1995) showed that skills networks could have different fractal levels in that, on closer inspection, a particular skill can prove to be a network of skills in itself.

People ask 'Can you drive a car?' They see the ability as a unitary lifestyle skill along with such things as shopping, banking and using the internet. When the skill of driving a car is examined closely, however, it can be seen to consist of a network of skills that are linked sequentially. First there is being able to manipulate the controls of a car, then being able to do so while the vehicle is in motion on a motorway infrastructure, then being able to do so in traffic, in accordance with highway regulations while being able to recognize, read and react to signage systems and, when necessary, call on and interact with auxiliary transport services. All these skills are critical and are tested before driving licences are issued.

The skill of driving is in fact what links the different NETS together on road transport. This is also true of the skills of driving trains, piloting planes and navigating ships. It is the skills network that is the integrating intelligence in NETS and it exists in two dimensions: the arbitrary knowledge of how to do something and its actual application. Throughout the history of transport it has been humans who could slip back and forth between these dimensions by knowing what to do and doing it.

If any of the critical skills that make up driving are in turn examined, then another skills level emerges. Being able to control a car means being able to steer, brake, accelerate, select gears, use indicators and so on, and these skills can in turn be broken down into elementary 'if–then' networks that a computerized system can control. Skills analysis, as it gets down to elementary network levels, provides the basis for automating the driving process. The human intelligence that physically makes transport happen can in the last analysis be replaced by AI.

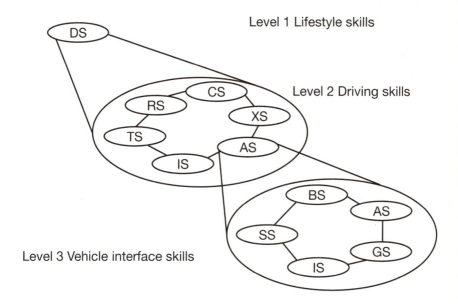

Figure 3.3   Fractal levels of skills networks

Level 1: The skill of driving (DS) is one of a number of lifestyle skills. Level 2: Driving skills are a network that links being able to interface with an automobile (AS), to doing so on an infrastructure (IS), in traffic (TS), according to regulations (RS), while being able to interpret signage and signals and road related communications (CS) and interacting as appropriate with auxiliary services (XS). Level 3: Interfacing with an automobile is a network of skills that include steering (SS), braking (BS), accelerating (AS), selecting gears (GS) and indicating (IS).

Transport communications theory

We are not trees. We depend on transport to bring the means of life to us or take us to them. Nor are we in endless motion relative to our immediate surroundings. We need periods of stability in one place where the means of life can be stored and processed. Similarly, for our species to survive we must be able to communicate with each other across space, but not endlessly. We also need time to think and sleep. Transport and communications for humans are episodic and each episode has its beginnings and ends and brings about change even if it is only in time and place.

When two nodes are linked they form the basic unit of a network. To see transport and communications as multiples of linked nodes is to see them as networked activities. We may conceptualize networks in isolation in order to design and cost and build them, but in reality networks link to other networks and all transport and communications networks are ultimately linked.

Atoms and information are moved from one place to another in a systematic way for a purpose. Something in one place is needed in another and the movement must, therefore, result in some kind of change, some process takes place involving what is moved. In this are the seeds of new transport and communications episodes.

There are no single simple transport networks. Transport is made possible by the intersection and interaction of different kinds of networks: infrastructure, traffic, regulations, skills, auxiliary services and communications. Without any of the NETS transport would not happen in any systematic way. That transport does happen systematically is because of yet another kind of network, that of transport services providers who use logistics to translate the transport needs of people into seamless network activity. However, transport may also happen when individual transporters arbitrarily access the enabling transport networks, as when someone decides to go for a drive.

Understanding transport as a complex interaction of different kinds of networks at different fractal levels provides a basis for analysing transport systems with a view to controlling and improving them. Today this is done by humans. Human intelligence interprets the communications in transport and controls the different networks to make transport happen, but it may not be that way in the future when much of the communications in transport is providing the information needed to control transport via AI.

4

First order meaning: clear transport communications

The City of Ames has an inventory in excess of 8,000 signs including regulatory, warning, guidance and informational. The signs comply with the directions for standards developed by the Federal Highway Administration in the Manual for Uniform Traffic Control Devices (MUTCD). The MUTCD contains national design, application, placement standards, and guidance for all traffic control devices such as signs, signals, and pavement markings. These traffic control devices regulate, warn, and guide road users along highways and streets all across the U.S. Traffic control devices are important because they optimize traffic performance, promote uniformity nationwide, and help improve safety by reducing the number and severity of traffic crashes. Records are kept to know the type, location and installation date of each sign. (http://www.city.ames.ia.us/worksweb/trafficdept/traffic%20signs.htm)

Introduction

One of the different kinds of networks enabling transport systems (NETS) is that of first order communications. Chapters 1 and 2 explained the quantitative factors that transport and communications have in common, ie the physical moving, storing and processing of atoms and bits of information. This and the next chapter are about the qualitative aspects of transport communications. What is the meaning of communications in transport? All communications have two orders of meaning. This chapter is about the first order of meaning. This is the level of meaning sought by the kind of communications networks described in Chapter 3. It is the level of meaning intended in transport signage systems, where what is communicated should be clear and mean the same to everyone.

Every transport system has a code of conduct for traffic that is implemented through a signage system. People who drive transport vehicles are expected to know all the signs in the relevant traffic code and to be able to act accordingly. One can only regard with awe the intellectual capability of the good people of Ames who, as cited at the beginning of the chapter, are expected to recognize 8,000 signs if they drive a car.

First order signification

The study of signs in Europe is derived from the work of Ferdinand Saussure (1857–1913) and is known as semiology (Barthes, 1967). In the United States it derives from the work of Charles Peirce (1834–1914) and is known as semiotics (Ogden and Richards, 1923; Ogden *et al*, 1923).

In university departments concerned with such aspects of communications as literature and journalism, a way of thinking known as post-modernism has developed in recent years. It sees every message as having a different meaning for every individual who receives it. What we are looking at in this chapter is the antithesis of this view of communications. We are looking at communications that are intended to mean precisely the same thing to everyone. They are communications that seek to remove any kind of ambiguity. In the words of the city of Ames cited above, traffic signs seek to 'promote uniformity nationwide'. If the people who drive vehicles interpret the signs they encounter idiosyncratically, they become a public menace.

There is a paradox in this. One of the basic axioms of communications, according to Shannon (1948), is that perfect communications, in the sense of moving a message from one place to another so that it has exactly the same meaning at its destination as it had at its source, is impossible. Between the source and destination of a message there is always some noise.

Nobody gets killed if they read a book, or view a film differently from everyone else, but when drivers of different vehicles interpret traffic signs differently the consequences can be fatal. Although there can never be perfect communications in transport signage, those who implement signage seek to reduce the noise in communications as much as possible and to make messages as clear as they can to ensure that everyone who uses a transport system has a common understanding of the signage code.

The meaning of signs in a traffic code at the literal, defined, commonly understood level is known in communications studies as 'first order signification' (Barthes, 1967). One could equally say 'first order of meaning' or even 'first order communications'. It is taking signs at their face value to mean exactly what they are supposed to mean.

Codes and signs

A code is a set of signs. Strictly speaking, therefore, a language is a code and the words in it are signs. However, the complexity of languages allows ambiguity. In seeking clarity, transport systems have developed their own codes based on flags, lights, guns, horns, Morse and semaphore, in which there are restrictions on the number of signs, the combinations in which they can be used and their meanings.

A sign is something that stands for something other than itself that is called its referent. Sometimes emergency crews forget to remove a 'hazard ahead' sign after the problem it denotes has been dealt with. Motorists seeing the sign look for the problem that no longer exists. The sign does not have a referent and so it no longer has meaning.

A sign, like a coin, has two sides to it, but, unlike a coin, only one side is real, while the other is virtual. One is called the 'signifier' and the other is what is 'signified'. The signifier is the physical manifestation of a sign, such as a red traffic light, the word 'stop' written on a signpost or a warning whistle from a train. What is signified is the concept the signifier evokes in the minds of the people who see or hear the sign. Road users know that a red traffic light or a stop sign means that traffic must stop. They know that a 'Hazard ahead' sign means there is a hazard ahead and that is why it does not make sense if there is none.

Signs may look like what they stand for, in which case they are classified as iconic signs; they may have a totally arbitrary shape or

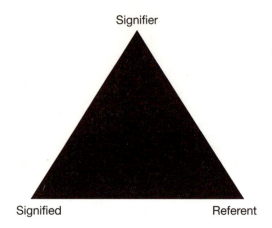

Figure 4.1   A sign has a signifier, a signified and a referent

sound, in which case they are classified as symbolic signs, or they may be linked in some way to what they stand for, in which case they are classified as indexical signs.

The meaning of a sign depends on the context in which it is presented. A traffic 'Stop' sign in a student's bedroom would not have the same meaning that it would have at a road intersection. Transport signs have first order signification in the context of the transport systems they are used in and from their association with other signs in the transport code they are part of.

In transport, signals are a special kind of sign that are actively trans-mitted to catch attention and require a specific action in response. At a railway crossing, a red light blinks, bells ring, the train gives a warning whistle and gates close. These are all signals that seek to catch the attention of road traffic and warn them not to cross the railway lines while the train is passing. Despite all this, level crossings take a steady toll of lives around the world.

Transport codes seek clarity, brevity and to be unambiguous. Official manuals or handbooks embody the communications rules and conventions that someone using a particular transport system should know. Sometimes, however, these are worded more to fit legal interpretations of their meaning than to communicate clearly to users. Anyone seeking proficiency in a transport system so that they can be certified as a driver, pilot or ship's master will need to know the codes of their transport system and will take examinations to establish that they do. These examinations will not test imaginative interpretation

Figure 4.2 An iconic sign

A sign is something that stands for something other than itself. This roadside sign is a 'signifier.' What is 'signified' is in the minds of the people seeing it, namely that there is a danger from falling rocks. The inset shows what the sign refers to (its 'referent'). The slope is steep and there has in fact been a recent fall. The image of rocks on the signifier is iconic because it resembles falling rocks; however, the inset shows what falling rocks actually look like. This is a standard sign used wherever there is a hazard of falling rocks. The signifier is a generic image that signifies to a rapidly approaching motorist, no matter what their language, that there is a danger of falling rocks. The distance between the sign and its referent is such that motorists can make the relationship and begin to take care.

of signs, but their precise standardized meaning and what action they call for.

A consequence of globalization is that many of the people using a transport system in one country may be tourists or visitors from another country speaking a different language. Iconic signs like that in Figure 4.2 transcend language and some signs have become universal. No matter what the country, red traffic lights mean stop. Increasingly, traffic signs are being globalized by using symbols that are standard worldwide and not dependent on language.

Figure 4.3 A symbolic sign

This sign stands for 'No parking' but bears no resemblance to such an action. The relation between signifier and what is signified is arbitrary. The motorist has shifted this moveable sign out of the demarked area to which it refers.

Information and redundancy

Shannon saw information as what was unexpected in a message and therefore the opposite to what was redundant because it was already known. For someone on their first flight, the explanation of safety procedures before takeoff is information. After that person has flown several times the repetitious parts become redundant. Most messages are a mixture of information and redundancy. A policeman waves your car down (the action is information because it is unexpected). He leans in and looks at you and your passengers and then says, 'This rain is forecast to get worse.' (This is redundant, you can see it is raining and you also listened to the forecast.) 'There is a bad crash up ahead and the road is blocked.' (This is information.) 'You have children with you.' (Redundant, you know you have children with you), but now the policeman says 'There is a turn-off a mile back and if you take it you will come to a motel where you might get accommodation'. (This is useful information. The previous seemingly redundant information

Figure 4.4 An indexical sign

The double lines in the centre are signifiers that signify 'Do not cross these lines.' The sign is part of what it refers to.

now takes on new meaning and provides a context for helping you decide what to do.)

An excess of unexpected information can be disturbing, an excess of redundant information is boring. Some repetition of information, especially when couched in a different way, helps to overcome distortions in a message that arise from noise. Transport signage seeks an optimal balance between information and redundancy.

Transport signage

The term 'signage' refers to all the fixed signs, as distinct from signals, that have grown up around a transport infrastructure. It includes signs erected by highway authorities, port authorities and airport authorities for the guidance of traffic. These may be compulsory as in the case of 'Stop' or 'Give way' signs, cautionary as in the case of signs that advise of wind gusts or the possible presence of animals, or informative as in

Figure 4.5 People waiting for their train

The passengers have tickets which tell them the time of departure, the platform and the destination of their train. Above them is a large display system which tells them the same thing and would seem to be redundant. However, this is where any new information about the status of their train could appear (if it is running late or if the time of departure or the platform has been changed). When the passengers go to get their train they will check from the sign at the end of the platform that it is the correct platform and the right train and that it is departing at the time given on the main notice board. The redundancy in these interlinked sign systems reassures, but how often will a passenger still feel the need to ask: 'Is this the such and such train for...?'

the case of signs that give directions and distances and details of where there are rest sites or refuelling facilities. Besides such official signs, signage includes the billboards, shop signs and advertisements that compete for the traveller's attention.

Particularly as it is found in downtown shopping areas or on city outskirts, the proliferation of such signage can cause confusion (Berger, 2005; Wenzel, 2005). Travellers become increasingly dependent on the signage that is part of the communications NETS for direction, warning, advice and guidance. Whereas at a conventional crossroads people have a common-sense feel for the direction they should take, at a clover leaf junction on a motorway the turnings they take may be counterintuitive and they are completely reliant on signage. The signs of shops, petrol stations and motels also have a utilitarian function to inform people of the ASPs that are available. However, these signs also have a competitive advertising function. Commercial signage adjacent to a highway is often restricted by highway authorities to minimize the confusion that comes from their dual function. Advertisements that clamour for attention distract from signage that seeks to control traffic.

Perception

Imagine driving along a motorway in the middle lane and looking for the sign for a particular turn-off. There is traffic close behind and ahead and a huge truck on your inside lane is travelling in such a way that it is between you and any roadside sign. Motorways usually allow for this problem and have more than one sign for an intersection, but there are many big trucks. It is not enough to put up a sign. People have to

Figure 4.6   Downtown signage in Shinjuko, Tokyo

The neon lights compete for the attention of people in the street and it is difficult to pick out the traffic lights.

be able to see it and read it in time to do something about it. The terms distal stimuli and proximal stimuli are used relative to each other to distinguish the place from which a message is transmitted to the place where it is received. These terms equate with Shannon's basic model of an act of communication having a source and destination.

Perception in humans begins when stimuli impinge on the receptor nerves of touch, taste, smell, sight and sound and ends when the stimuli are identified and something is signified. Transport is not concerned with smell and taste any more, though there was a time when fishermen could tell where they were from the taste and smell of the water. Touch is important at the micro levels of movement, as when we feel the reassuring grip of banisters, an elevator button, the kick of a

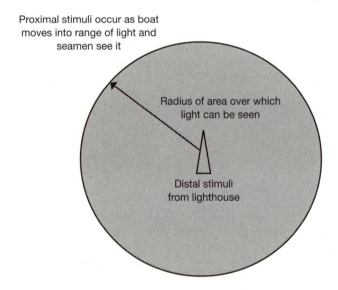

Figure 4.7　Distal and proximal stimuli

At night at sea in the days before radar and radio, sailors would anxiously scan the horizon for the signal from a particular lighthouse that would enable them to know their position. To sailors on lookout, perception could last several seconds as they tried to decide if they really did see a flash of light. They would look away and wipe their eyes or the glass of their binoculars. When they were convinced there was a light they would then have to decide if it was the light they are looking for. All over the world there were, and still are, lighthouses beaming distal stimuli to mariners looking for them, but the history of lighthouses is one of shipwrecks.

rudder and the controls of a car, but it is stimuli from sight and sound that are of particular importance to perception in transport today. Clear transport communications has to take into account variation between people's perception and how they relate to the special factors of a particular transport mode. The way people perceive signs when they are travelling at speed on a motorway is not the same as when they are walking.

In modern urban sites people are inundated with sensory stimuli. They cannot pay attention to them all. Perception has to be selective. It is not enough that they know the code of a particular transport system and what to look or listen for. The signage of that code needs to be made and placed so that it can easily be seen. People look and listen for transport signs as a hunter would, because survival depends on their ability to see traffic lights or that the car in front is turning. When there is something new, unexpected and dangerous, signs must grab people's attention. To do this signs may be big, frequent, loud, intense, intermittent and moving.

Signage heuristics

We cannot claim to understand perception in humans in the way we understand how computer-recognition of visual and audio stimuli works. Traffic signage is essentially based on heuristics. These are rules of thumb that come from experience and common sense. For example, since transport signage has the purposes of regulating, warning, guiding and advising people who are part of the traffic, it makes sense that all transport signs should:

- be recognizable in relation to the sensible speed of traffic;
- convey well-defined, easy-to-understand messages that clearly denote what the signs mean;
- not be lost among other signage or confusing stimuli;
- provide enough time for the driver to read and respond;
- conform to the context they are in and to any other signs with which they are linked;
- conform to the standards that define their selection, placement and usage;
- complement and not contradict each other;
- provide an optimal balance of information and redundancy.

As computing capability has grown over the last half-century, it has become possible to conduct empirical research to substantiate and refine or reject the heuristics on which transport signage is based. However, one of the main causes of accidents in transport remains a failure by people to observe transport signage and traffic signals in the way intended.

Traffic flows in transport systems according to signs that are based on rules which may be anything from local ordinances to national and international law. It is far easier to add new regulations than it is to remove old ones. So rules and regulations grow over time to create the kind of complex highway code that we see in the city of Ames. They enmesh transport systems in a regulatory network that makes change difficult and may contribute to gridlock rather than resolving it.

Wayfinding behaviour

Unique to transport is a form of communication described as 'wayfinding' (Golledge, 1999). To make a journey of any complexity requires an ability to conceptualize the route as a whole and in part. The road that people see in front of them is indexically linked to the larger network of roads that they cannot see. They need to be able to link the signs in their immediate transport vicinity with the transport situations that are imminent and with the route they are taking as a whole.

Experiments with rats in mazes and studies of migrating birds and flying insects seem to show that they can develop mental cognitive maps to which they can relate direct observation of landmarks and celestial features. We can recognize in ourselves an ability to recall a route we have already traversed. What seems unique to humans, however, is that they can also visualize a route they have never traversed before. They can do this from a verbal description or a map. Maps have been used to enable transport systems in many cultures for thousands of years and today maps are readily available at every level of transport. We are, however, moving into a new era of 'wayfinding' behaviour based on global positioning systems (GPS) equipment and radio frequency identification (RFID) tags. With mobile transceivers that link to GPS satellites it becomes possible for people to see where they are, where they might want to go and how they can get there. With RFID coupled with GPS it is possible to check the location of any plane or boat or indeed of any person or animal or item of freight.

If all automobiles were fitted with RFID it would be possible for a centralized computer system to map the total traffic network on a given road infrastructure network and control it to optimize the performance of traffic as a whole, which is how aviation and marine traffic networks are becoming organized.

Signage and IT

Transport signage systems work most of the time, but not all the time. Motor transport kills far more people than terrorists because of human error and, in particular, failures to follow the highway code. Despite the care with which signs are designed and placed, people often do not perceive them or interpret them as they are supposed to, according to law and first order signification.

People get impatient, frustrated, angry, excited, hyperactive, drowsy and depressed. They sometimes drink too much and no two people will ever see the same sign in exactly the same way. What is more, despite all the warnings, road safety campaigns and police traps, people will continue to behave like this because that is their nature. Computers do not have these human behavioural problems. Computers only process information at the first order of meaning. If they detect a sign then they do what the sign expects them to do. A car driven by a computer recognizing a sign that signified 'Stop' would stop. It would not have a set of qualifying thoughts along the lines of 'There is no one about to see and I am in a hurry so coming to a complete halt is a bit childish'. The cybernetic interaction between traffic, and between traffic and its infrastructure, depends on first order meaning which is what computers understand. So why not use computers instead of humans to drive vehicles?

We move in this direction. Planes and ships use automatic pilots linked to traffic control systems. The interaction between high-speed trains and their infrastructure is computerized. Transport systems increasingly have built into them single purpose computers linked to sensors that continually check on performance and behaviour. Modern motorized vehicles can advise that they need fuel or a check up, will stay below the speed limit and will not move until the driver has fastened his or her seat belt.

The trend towards using IT in transport is restricted, because the networks of first order communications transmit their messages by light and sound in channels intended for human eyes and ears in ways that, as yet, computers cannot match. Computers can remember better than people, do maths and play chess at a level people can no longer match, but as yet they cannot see or hear like people. Sensors have long been able to detect an oncoming light beam and dip a car's beam accordingly, but as yet computers do not read road signs. The complexities of downtown traffic environments and signage are beyond workaday computing capability in the early years of the third millennium. But this will not always be the case. Signage systems will evolve on roads, as they are doing on rail and by air and water, to an electronic interchange by telecommunications between the computers and sensors within vehicles, between vehicles and between vehicles and the infrastructure. The communications NETS of the future may well be unintelligible to people and so avoid the problems created by human ability to understand communications at the second level.

5

Second order meaning: fantastic transport communications

Fire up the Jaguar with the bright red button on the console and there is a dramatic little tremolo from the starter motor before the V8 settles into a woofling idle. Move off and there is a low rumble that turns heads on the pavements, as though the ghost of Steve McQueen was cruising by. And on the open road in the brief interval between flooring the accelerator and breaking the law, there is a glorious bellicose blast that if bottled could be sold as Essence of Motorsport. (R Wilson published in *The Australian*, 5 July 2006)

Introduction

The relationship between signifier and signified is called signification. When we perceive the signifier of a sign and know what it signifies we think we know what it means. Or do we? The French philosopher Roland Barthes (1977) made the case that there are two levels of meaning in signs: a first order of signification where the meaning is the obvious literal one of what a sign denotes and a second order of signification in what a sign connotes.

The traveller was met at the airport by a professor of philosophy who helped him carry his baggage to an old Volkswagen. The two chatted amiably about the forthcoming conference as they drove out of the airport. They came to some traffic lights which were at red but to the traveller's surprise the professor blithely drove straight through them. Fortunately it was early morning and there was no other traffic. They came to another set of traffic lights which this time were green and the professor stopped. Completely baffled and wondering whether he had come to the only country in the world where the meaning of traffic lights was reversed, the traveller asked the professor what his philosophy of traffic lights was. The professor responded 'Rules are for the guidance of the wise and the obedience of fools. If I see no traffic coming why should I stop at the command of a machine?' 'Indeed' said the traveller 'but why stop at a green?' 'Ah' said the professor 'that is in case some idiot is shooting the red'. *(Travellers' Tales)*

The professor in the story above knows that the first order signification of red traffic lights is stop and of green traffic lights, go. But he also thinks that the requirement to obey traffic lights is a sign of growing domination of people by machines. This for him is second order signification.

The combination of computers and telecommunications, referred to as IT, can parallel human communications in its ability to transmit, store and process information. The last chapter concluded by noting that IT has the potential to conduct the NETS communications that are critical to transport and to do so more effectively and safely than humans. IT communications grows to rival, and in time could surpass, human communications in its ability to conduct communications *at the first order of meaning*. Unlike humans, all the computers in a given system can be programmed to recognize the same signal in the same way and react in the same manner. As yet, however, computers do not have imaginations and show no signs of matching humans at the second order of signification, which is what this chapter is about.

Second order signification

All signs carry the two orders of meaning. Which of the two orders of meaning is the one that is intended in any particular communication

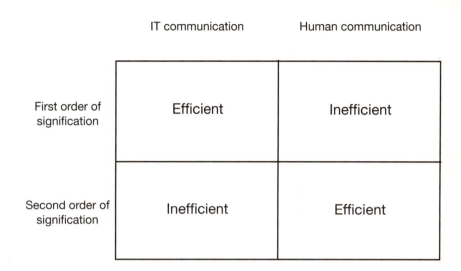

Figure 5.1 Relative effectiveness of IT and human communications at the first and second orders of signification

depends on the purpose. Chapter 4 used the example of transport signage to illustrate communications where the intention is to prioritize first order signification. People may harbour their own strange thoughts about traffic signs, but to drive safely they know to interpret signage at the first order of signification and that this will be reinforced by traffic police. An example of communication that prioritizes second order signification is the review of a new car quoted above. It seeks to sell the model, not by explaining what a purchaser will get for what price, but by encouraging people to think that owning such a car imparts the panache of a glamorous film star. Second order signification in transport communications comes into its own in advertising.

Critical theory

Barthes (1995) saw second order meaning in communications as a way of linking people with the dominant ideologies of a society. This is based on the Marxist idea that our sense of who we are depends on the society we are in (Marx, 1859, 1977). From this it follows that the dominant group in a society can, through their control of mass media, establish hegemony over the rest of society by getting them to accept and conform to the dominant ideology (Forgacs, 2000). Hence, the Orwellian image of fascist dictatorships using mass media to instil obedience.

After the Berlin wall came down, the Marxist tradition in the study of communications lost some of its political and economic connotations, but remained an influence in a branch of communications studies known as critical theory. The term reflects a broad concern for the inequalities that arise from the way communications systems are managed. Feminist studies is a branch of critical theory that argues that men have used their control of mass media to establish a male hegemony over females by subtle second order signification in communications. For example, until recently the advertising of cars, in the manner of the Jaguar review cited, would portray drivers as Steve McQueen-style he-men. Women were seen as admiring passengers.

According to critical theory, transport systems in search of profit disregard ecological hazards, cultural issues and the aspirations of developing countries. The theory has expression in the anti-globalization movement and tends to view the development of global transport systems critically, seeing them as tools of oppression at work on a global scale (Eschle, 2005).

Symbolic interactionism

An alternative field of thought based on the ideas of Sigmund Freud is that second order signification is linked to the subconscious (Mitchell and Black, 1996). In other words, second order signification has its roots in the self, rather than in society. At the first order of signification messages are intended to mean the same to everyone. At the second order of signification everyone interprets a message differently. Psychology, especially the psychology of perception, has played an important role in trying to understand and resolve the irrational actions of humans in their interactions with transport. This has led to the development of two specialized fields of study in transport: traffic psychology, which is the study of the behaviour of road users (Rothengatter, 1997) and transportation psychology which is concerned with human motivation in transport (Rothengatter and Huguenin, 2004).

In the United States an interesting approach to communications that comes somewhere in between the extremes of Marxism and Freudianism is symbolic interactionism (Charon, 1992). Like Marxism this theory sees individual consciousness as being based on social interaction, but unlike Marxism it sees social interaction as depending upon the unique capability of humans to be conscious of themselves and conscious of how they are seen by others. For example, in making the decision to turn at an intersection we take into consideration the traffic around us. We consider the consequences of our actions from the point of view of other road users as well as ourselves. This imaginative extension of perception to see ourselves as others see us is a second order of signification in contrast to the first order of signification that comes from direct perception of what is in front of us.

Symbolic interactionism is associated with a number of similar theoretical approaches which see communications as having a second level of meaning in which people think of themselves as involved with others as part of an act (Hickman and Kuhn, 1956), a drama (Goffman, 1974), a narrative (Fisher, 1970), or a fantasy (Bormann, 1985). Behind all this theory is the idea that people turn the events around them into some kind of story or theatre in which they are both actors and audience.

Theories such as these are difficult to prove because we do not know what goes on in other people's heads. It is through our own experiences that we find ourselves in agreement or disagreement with them. Does our engagement with traffic seem like theatre? When we are engaged in some kind of transport activity do we feel that we are part of some ongoing story that has its heroes and villains and in which we have some special role to play? Acts in transport may be small private events, but according to these theories, they accumulate and merge to become the myths to which societies and cultures subscribe. How often when we turn to the news are transport events the big stories? We romanticize transport. It is one of the principal themes of books, films and television. Travel suggests excitement and adventure, going forth to brave the world and find new places and faces and coming home to loved ones and security. War is all about the transport of men and missiles to communicate hate and win glory and many stories finish with a journey that ends in lovers' meetings.

Marxist theory sees mass media as a vehicle for imposing a world view on the masses. According to this theory, people buy transport because they are the victims of mass marketing. US theories of mass communications, in contrast, see the media as attending to people's needs. What we get in the media is what we want. We want cars and car advertisements are there to help us get them and so that we can participate in the stories that go with them. This is needs and gratifications theory according to which it is not communications that shape people but people who shape communications (Katz, Blumler and Gurevitch, 1974).

Figure 5.2 *The Sultana* (1860) Artist unknown

The painting denotes the barque Sultana taking in sail off the cliffs of Dover. Connotation depends on the background of viewers and the times they live in. A sailor would recognize that, with the cliffs of Dover on her port side, the Sultana is heading up the English Channel and so is homeward bound, with all the connotations this has of journey's end, success and achievement. This painting was in fact commissioned by the ship's owners, who intended to advertise the ship as the latest thing in transport: she was a large barque that could be handled by a small crew and so compete with the new steamers. Today such a picture might connote the possibilities of using renewable energy.

Cognitive dissonance

We seek to make the symbolic world in our minds consistent with the physical world we live in. We may even, for a while, succeed in deluding ourselves that they are. But when the heroes and villains and objects and settings of our fantasy life do not behave in real life in accordance with their fantasy roles, then we suffer from what Festinger called 'cognitive dissonance'(Festinger, 1957). According to this theory, people seek consistency. They organize their mental world so that the things within it are consonant with each other and the world without. If reality intrudes with things that are dissonant, then people get disturbed. They seek to resolve this dissonance either by changing the real world so that it matches their mental world or vice versa.

The design of transport systems is an endless attempt to resolve the cognitive dissonance that inevitably occurs in transport because there is always some kind of 'noise' involved. Advertisers sell glossy images of dreamy air travel with smiling people enjoying wonderful food served by attentive staff on smart aeroplanes that are always on time. Some airlines manage to some extent to match the reality of their flights with the promise of their adverts, but only in first class. Too often today, the reality is one of crowded airports, overworked staff holding on to frayed tempers and queues for toilets and security checks. This produces cognitive dissonance which mounts if not resolved. There have been riots in airports and if the grounds for cognitive dissonance increase there will be more.

Figure 5.3 Concorde

Concorde was never a commercial success, but was regarded as one of the most beautiful aeroplanes ever built.

Figure 5.4 The Bugatti Royale

The Bugatti Royale is held to be the most beautiful automobile ever designed. Only six of them were ever manufactured.

Figure 5.5 The *Normandie*

Built in St Nazaire, France in 1912, the *Normandie* was then the world's largest and fastest liner. It was intended to be a floating showcase for French haute couture and was an extremely beautiful ship.

Figure 5.6 Valencia railway station

It looks like the entrance to an exotic theatre. Enter here for a magical mystery tour.

Figure 5.7 A modern airport lounge

Strictly utilitarian, there is little concession to aesthetics. This is simply a place to sit and wait.

The aesthetics of transport communications

An area of transport communications that seeks a balance between the first and second orders of meaning in communications is in the design of transport infrastructures and vehicles. Besides a utilitarian concern to provide efficient and cost-effective transport systems, there is sometimes also a concern for the aesthetics of the transport experience, for the beauty of the scenery passed through, the looks of the vehicles used and the quality of the service provided.

The graceful lines of an aeroplane, a ship or a car derive from aerodynamic design that seeks to improve the speed and energy efficiency of getting from A to B, while at the same time pleasing the eye in the way it suggests the nature of speed. The curves of a motorway may add a little to its length, but in compensation may harmonize the road with the curve of the countryside in a way that offers travellers sweeping vistas and keeps drivers from getting bored and drowsy.

First- and second-order meaning in organizational communications

Transport, like any other systematic activity, depends on organizational communications. Every one of the NETS is held together by its organizational communications. They too seek a balance between first and second order communications and it is not always easy to find. Managers want their orders and instructions to be unambiguous. They may also want to inspire team spirit and worker participation. The two are not necessarily compatible. If managers are successful they are seen as leaders, if not, as autocratic tyrants. Organizations may seem to be held together by the hierarchical webs of the organograms that define how first order corporate communications should proceed, but they also have unofficial grapevines where second order communications is dominant and gossip can build or destroy corporate culture (Kreps, 1990).

A lot of organizational communications is conducted face to face. But face-to-face communication is not only in the words people say, it also includes the intonation they use, their posture, body language, dress and grooming. These are all vehicles for second order meaning that can contradict or reinforce what the spoken words denote. To reduce ambiguity, official communications can be conducted by writing, but this medium lacks flexibility. It is difficult to change or erase written words, especially now that they derive from computer systems with memories that last for ever.

For humans there is always a second order of meaning in a message and always the possibility that this will influence instructions. Where organizational communications seek complete compliance it would seem logical to suppose that they will increasingly use computerized communications. A very real problem is where organizations seek to get people to behave as though they were computers and did not have second order thoughts.

IT and second order signification

> Australia is famous for its remote and lonely pubs. Into one of them a few years ago came the traveller tired and weary. Silent locals stared at him. A bored barman asked 'What'll it be?' In front of the traveller floated an advert that read 'Drink Smirnov and get into virtual reality'. 'I'll have one of those' he said. The barman gave him his glass of vodka and a ticket. 'What's this for?' he asked the barman. 'The virtual reality you get into when you've drunk that vodka' replied the barman and he nodded 'behind you'. The traveller turned and there was a beaming promotions person standing beside a strange barrel-shaped device. 'Come sir,' she said 'just step into this machine. It is perfectly safe.' She put a pistol in his hand. 'Don't worry' she said 'just shoot anything that tries to get you.' Then she put a helmet over his head and he found himself immersed in a virtual reality game where a sinister figure was shooting at him and pterodactyls kept flying down to attack him and he was blazing away with his gun, fighting for his life. Suddenly he missed and a flying monster got him and he was back in the pub in the middle of nowhere. But things had changed while he was in virtual reality. The locals had been watching to see what happened to the first person to try the device and now they were quaffing vodka and queuing up to get into virtual reality. (*Travellers' Tales*).

The matrix in Figure 5.1 shows humans as being more expert than IT at communications that prioritize second order meaning, but will that always be the case? The story above is true and has a moral. If you don't like it where you are, get into virtual reality. This is in fact what we do when in some dreary waiting lounge we open a book or a magazine or watch a video or simply daydream about the travel adventure we think we are having. People now live so much in the fantasy world provided by the media that transport systems have to cater to this. Aeroplanes have in-flight video services. Stations and airports have video game areas. People walk around with their personal audio players.

> The traveller in Australia went to see the Great Barrier Reef and found himself sitting in the stern of a luxury boat as it powered out to the reef. With the trade winds in his hair and the sun glancing off the sea he wondered where all the other passengers could be. He eventually found them sitting in a large air-conditioned cabin watching a video of the boat they were on as it powered out to the reef on a beautiful sunny day with the trade winds blowing. (*Travellers' Tales*)

How often do the great jets have the shades down on their windows so that their passengers can watch the movies as they fly across the Himalayas, the Andes, the Alps and the wastes of Greenland and Siberia, blind to the vast tableaux below them? In the future, when we are travelling we may find ourselves less and less involved in the reality of the transport process and more and more passing the time in media realities. The media for fantasies improve in the sense that they become more and more realistic, so much so that what they portray could come to seem more real than what is real. In which case, will we any longer need to travel to realize our fantasies?

6

Transport paradigms and the episteme of globalization

A paradigm, in a sense, tells you that there is a game, what the game is, and how to play it successfully. The sense of game is a very appropriate metaphor for paradigms because it reflects the need for borders and directions on how to perform in order to 'do it right'. A paradigm tells you how to play the game according to the rules. (Barker, 1993)

... the main aim of [globalization] is to maximize the volume of global trade. (Candemir, 2004)

Introduction

This book is about the way causal circularity in transport and communications leads to globalization. The previous chapters have addressed the nature of transport and communications. This chapter addresses the nature of globalization.

Candemir (quoted above) expresses the way most people see globalization: as the process of increasing global trade made possible by advances in transport and communications. We might suppose, therefore, that globalization began when Magellan's ship completed the first circumnavigation and the traders of Europe began to think about what it meant for them. In 1776, Adam Smith published *The Wealth of Nations* which was to capitalism what Karl Marx's *Das Kapital* was to communism. This work provided the logic of free trade and a bible for globalization. The 18th and 19th centuries saw the technologies of trade and communications improve and a consequent expansion of world trade. Countries that did not subscribe to the principles of Adam Smith were colonized by those that did.

The first half of the 20th century, however, saw a hiatus in the growth of globalization. The two World Wars, the great depression and the rise of protectionism held back world trade. It was not until the second half of the 20th century that globalization, powered by new transport and telecommunications technologies, took off again. However, the growth of free trade remained constricted by the fact that much of the world was communist, socialist or had some form of state welfare.

In 1989, the Berlin Wall collapsed and so did the Soviet Union and the communist countries that were part of its empire. With the end of the Cold War, capitalism was the only game in the global village and global trade was unfettered.

It is from this time that people have become increasingly conscious of globalization. Gathering momentum, it has become a force in itself. Globalization is not just an economic environment, it is also a state of mind, a way people see themselves as citizens of a world in which they trade the same things and watch the same television programmes. There is a sense of global community in the way people respond when there are tsunamis, earthquakes and storms and share concern about global warming and avian flu. Everyone listens to radio, even the poorest. Most people watch television, even if it is not their own. The people of the world begin to see and hear each other and to think about the same things, but not in the same way. People become aware how rich or poor they are relative to others and whether in the words attributed to Nicholas Negroponte they are 'part of the steamroller or part of the road' (Brand, 1987).

Such an all-encompassing shared world view has been termed an 'episteme' by the French philosopher Michel Foucault (1972). An episteme is a paradigm that encompasses all other paradigms. To understand an episteme we first need to know what a paradigm is.

Paradigm

The term 'paradigm' was popularized by Thomas Kuhn (1922–1996). Kuhn (1962) applied the term to science as it exists in the minds of all the people who think of themselves as scientists and in all the books and articles that scientists agree are scientific in nature. According to Kuhn science is what scientists think it is. This has been taken to imply that science is more a culture to which scientists conform than a search to understand the nature of reality. Following Kuhn, the term paradigm has been applied to such fields as education (Heinich, 1970), technology (Perez, 1983) and universities (Tiffin and Rajasingham, 2003). It is now widely used in the context of any organized body of knowledge and that includes transport studies and communications studies.

In the Kuhnian sense a paradigm is knowledge in the abstract organized as a system. People who *know* a body of *knowledge* in its paradigmatic form *know how to* do something with it in a way that conforms to the paradigm. In this sense, a paradigm is like a game, as Barker (cited at the beginning of this chapter) would have it.

The term paradigm is also used in linguistics where it refers to language as a body of knowledge that exists in the abstract as distinct from its actual manifestation in speech or writing which is known as a syntagm. By the same token, if knowledge of how to play a game is a paradigm, then actually playing a game according to the rules of the game is a syntagm. In aviation the knowledge needed to pilot a plane is a paradigm and an actual flight is a syntagm of that paradigm.

The importance of the relationship between paradigm and syntagm is in the cybernetic adjustments that take place between them, particularly at the second order of cybernetics (not to be confused with second order meaning discussed in Chapter 5). Every syntagm of a paradigm provides feedback to a paradigm as to how effective the paradigm is. The paradigm then cybernetically adjusts. In this way paradigms change as they are used. When someone speaks a sentence this changes the language paradigm because it adds, however slightly, to the frequency with which certain words or expressions or sentence constructions are used. Paradigms and syntagms are locked together in circular causality.

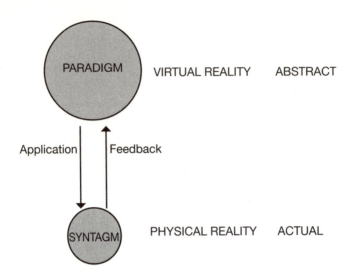

Figure 6.1 The relation between paradigm and syntagm

Paradigms exist in the abstract as knowledge shared by its practitioners. An application of a paradigm is a syntagm which is manifest in physical reality. Every syntagm of a paradigm provides feedback to a paradigm as to how effective the paradigm was. The paradigm then cybernetically adjusts.

Transport paradigms and syntagms

Driving a car is like speaking a language. It is a syntagm of the driving paradigm. Drivers read a road and drive along it according to the grammar of the relevant highway code. Driving a car today is different from driving one 100 years ago. Just as the use of language cybernetically changes the language paradigm, so the way people drive changes the driving paradigm which then changes the way people drive. However, it is not only the way people drive that changes the driving paradigm; it also changes because cars change, roads change, regulations change, signage changes and the services provided for transport change. The different networks that enable transport systems (NETS) all have paradigmatic and syntagmatic dimensions that have to be compatible with each other. Infrastructural paradigms must be compatible with traffic paradigms, regulatory paradigms, communications paradigms, service paradigms and skills paradigms. If one of the NETS changes, the rest of them must change to adapt. The NETS themselves are paradigms that have paradigms nested in them. The railway infrastructure paradigm includes paradigms for bridges, cuttings and stations and the railway station paradigm includes paradigms for ticket booking, platforms and parking. Changes in these subordinate paradigms will in turn effect change in associated paradigms.

The transport studies paradigm

Education in the broadest sense of the word is a process of learning the many paradigms that make up an episteme. People go to university to learn professional paradigms such as medicine so that they can

Figure 6.2   Syntagms of the Spanish railway train paradigm in 2003

The trains all look alike and conform to the current paradigm among railway engineers in Spain as to what a train should be like, but there are minor differences between these trains. Changes are made to improve them from year to year. A photograph of Spanish trains 50 years earlier would be different. We could perhaps think of the Spanish railway train paradigm as being something of a dialect in the broader global technological paradigm of railway trains of which all trains can be considered syntagms.

Figure 6.3　A syntagm from the *Shinkansen* railway train paradigm

The famous bullet trains of Japan are so distinctive in appearance and have so many special features that they could be thought of as having their own paradigm within the Japanese railway train paradigm within the global railway paradigm. Think of how the English language varies from country to country and within countries.

syntagmatically apply it to patients and the paradigm of law so that they can syntagmatically apply it as practising lawyers. Is there a transport studies paradigm? Can we say that transport studies now takes its place alongside engineering, medicine and law as a professional field of study whose graduands go forth to practise transport as doctors practise medicine or lawyers practise law? A paradigm has to have a syntagmatic dimension. It is not knowledge for the sake of knowledge. It is knowing something in order to be able to do something. What do people who do transport studies know how to do in consequence of what they learn?

Like 'communications studies' the term 'transport studies' suggests we are still seeking to understand what it is. Both transport and communications lack a core body of theory and practice. The quotation at the start of Chapter 1 from the manifesto of the Transport Communications International Union provides a historical summary of the way different groups of people with different functions in transport have gradually found common ground. They are seeking to establish a paradigm. However, people in the transport industry still tend to think paradigmatically from within the shipping, aviation, road and railway paradigms of their countries where they are clear about what they know, what they can do and who their peers are. It is as we move towards integrating these paradigms into a larger global paradigm that the need for an overarching transport studies paradigm becomes obvious.

Epistemes

Transport paradigms are linked to other transport paradigms. They are also enmeshed in and linked to paradigms outside transport such as those of law, language, education and clothing. Try boarding an aeroplane in your pyjamas, or using the metro in a country whose language you do not speak. The term episteme refers to the totality of intermeshed paradigms in a society and the way this gives rise to a common way of thinking among people. To Foucault an episteme was a supra-paradigm of all the intertwined concordant paradigms by which people thought and behaved. Foucault believed that the world view from within an episteme was so encompassing and interdependent that it was impossible to think outside it (Foucault, 1973).

Paradigm shifts and epistemic shifts

Kuhn popularized the notion of paradigm shifts when he argued that the main changes in science were not in the gradual modification of the normal science paradigm through its syntagmatic application to real world phenomena, but in revolutionary changes that occur from time to time and radically shift the scientific paradigm. Such paradigm shifts are brought about by the discovery of some paradox that the normal paradigm cannot resolve. This then leads to the development of a new paradigm that does resolve it. Kuhn was thinking of the scientific revolutions brought about by people like Galileo and Newton. We can see the same kind of thing happening in transport and communications, not so much from recognition of a paradox, as from an invention that radically changes the way of doing something. Stevenson and Brunel are the Newtons and Galileos of transport and Marconi and Bell had a similar revolutionary role in communications.

Since paradigms are interconnected, a change in one paradigm will impact on other paradigms and require a process of adjustment. Radical change in a paradigm can have a domino effect. A paradigm shift or a conjunction of paradigm shifts can lead to a shift in the episteme. Below we look at how paradigm shifts in communications and transport technologies brought about the epistemic shifts we call the Industrial Revolution and the Information Revolution. However, epistemic shifts may also come from within a state by revolution or from without by invasion, conquest and colonization.

Empires are epistemes in expansion. To extend their territory they need good transport and communications. The Roman Empire was built by roads, the US and Russian Empires by railways, the British Empire by seaways. Imperial expansion is also marked by war and war is conducted with specialized transport systems. Victory is credited to the generals, but as often as not, it is because one side has a superior transport or communications paradigm.

The Middle Ages saw the rise of a feudal episteme in Europe based on the transport paradigm of knights in armour on horses. This involved specially bred war horses, controlled by curb bits, that could carry the weight of heavily armed knights; war saddles that provided a secure seat so that knights could apply force to the tip of their lances and iron stirrups that allowed them to stand and hack down with their swords on foot soldiers. Knights were the medieval shock troops that dominated in battle. They were of special value to the kings and lords of Europe who ensured their services with grants of land (White, 1966). The knight on horseback paradigm was not a single dramatic invention that radically changed the way of war, but a gradual improvement and inter-adjustment over almost a thousand years of the subordinate paradigms that make up the norm paradigm of cavalry. What provided the paradigm shift that ultimately unseated the knights in warfare was guns. Other examples of paradigm shifts in transport communications that have given an edge to empire building are Napoleon's use of a semaphore system to give him control of Europe, the British fleet's system of signaling with flags that gave it advantage at sea and the way European empires were extended by shifting the epistemes of tribal people with gunpowder.

In the 1860s people investing in transport could either bank on stagecoaches or railways, sailing ships or steamships. Anyone setting up an aviation system in the 1920s had a choice between Zeppelins and aeroplanes, in the 1970s it was between jumbo jets and Concorde. Those who are responsible for planning, designing and investing in transport developments have at times to decide whether to continue with the conventional way and invest in the norm paradigm or recognize that there is a paradigm shift and adopt and invest in the new ways. If the latter, they may well have to choose between paradigms that are competing to be the new paradigm, as was the case between Zeppelins and aeroplanes. Such decisions are a gamble. A useful guide is to look at past form.

Paradigm shifts in transport at the site level (1810–2010)

Sites are the places where people live, play and work. They are primarily buildings but include such things as docks, sports fields and mines. At this level the movement of people horizontally over short distances has changed little over the last 200 years. People still walk, lift and carry much as they have always done. What change there has been in horizontal people movement in buildings has come with the growing use of travelators and wheelchairs. There has been more innovation in vertical movement. Otis demonstrated the first safety elevator in 1852 and from the 1860s inner-city buildings have been growing upwards.

Where we can clearly see paradigm shifts at the site level is in the handling of freight in bulk. The first paradigm shift came with the widespread introduction of steam-powered derricks in the 1860s to replace block and tackle and pulleys. The second was in the 1960s with the introduction of containers (Rimmer, 2003). This was a paradigm shift that ran through all levels of goods transporting, but it was the on-site ability in dockyards and goods yards to load, unload and sort cargo in containers which dramatically reduced cargo handling and turn-around times, and made inter-modality in transport possible.

Paradigm shifts in transport at the urban level (1810–2010)

In the period up to 1860 urban transport was powered by people and horses. Poor people walked, carried, pushed and pulled, while wealthy people rode horses or were carried. For those who were neither rich nor poor, in some cities and towns there were urban tramways with horse-drawn carriages. Heavy goods were handled by horse and cart.

People still walk in urban areas. Infrastructures of pavements, pedestrian precincts, pedestrian bridges and underground sewage have improved the way they do this. The development of the bicycle in the later part of the 19th century and more recently skates and skateboards means that in urban transport people-powered transport continues to have a role. However, the use of animals in urban transport has declined. In developed countries they seldom appear in urban transport except for crowd control, state funerals and trips round a park.

Urban mass transit systems, at first powered by steam, began in the 1860s with the London Underground. This decade also saw the coming of steam-driven railways for commuting and electric trams. The 1960s

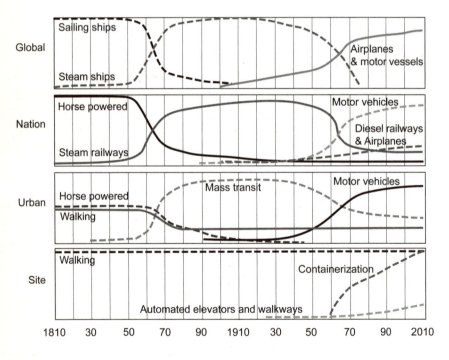

Figure 6.4 Paradigm shifts in transport from 1810–2010 at four fractal levels

The four levels are: the global level where the nodes are countries, the national level where the nodes are towns and cities, the urban level where the nodes are buildings and the site level where the nodes are rooms.

saw mass transit shift to oil for power. Trams were largely replaced by buses and the electrical energy that drove underground railways was increasingly generated by oil rather than coal.

In the early 1900s motorized vehicles appeared and after the First World War a growing network of petrol stations and garages extended their use not only within urban areas, but between urban and suburban areas where they competed with commuter mass transit systems. By the 1960s privately owned motorized vehicles were rivalling mass transit systems in urban transport and wherever urban governments have not imposed restrictions, motorized transport has become the dominant mode of urban transport.

Paradigm shifts in transport at the national level (1810–2010)

Interurban transport was primarily by staged horse-drawn vehicles until the 1860s when railways became the dominant mode and provided the transport and communications infrastructure that held nation states together. Railways switched to diesel power in the 1960s and continue to be important, particularly for heavy goods. However, developed countries switched to motorways after the Second World War and state highways have become the dominant interurban transport infrastructure for goods and people. Domestic aviation began around 1930 and since the 1960s has grown to rival railways for passenger traffic in developed countries.

When speed was not a consideration, inland waterways and coastal shipping lanes have provided the cheapest and most efficient modes of transport for freight at the national level and between neighbouring countries that share access to waterways. The infrastructure of rivers, lakes and coastline is a given that engineers have sought over the centuries to improve with locks, canalization and dredging. Waterway transport underwent a parallel change with other transport systems in the way sail on open waters and horses and people on towpaths gave way to steam engines in the 1860s, which in turn gave way to diesel engines in the 1960s. However, the shape of boat hulls, the speed at which they can go and the amount they can carry has not changed to the extent that stage coaches did when they gave way to railway coaches.

Paradigm shifts in transport at the global level (1810–2010)

Sailing ships gave way to steamships in the 1860s. Ship movements then became predictable and their trade routes independent of wind systems. But the smoke steamships made was visible for a long way and led to enormous losses from submarine attacks in the Second World War. Where merchant shipping was largely powered by steam at the beginning of the war, by its end all new shipping was oil driven.

The end of the Second World War saw a rapid increase in aviation for peaceful purposes and in the 1960s a paradigm shift with the introduction of jet engines. Flying became an economical way of travelling long distances and jet planes could fly above the worst weather so that air sickness virtually disappeared. The jumbo jet replaced the passenger liner as the main means of international passenger transport. The paradigm shift in aviation in the 1960s meant that today heavy cargo goes by sea and passengers by air.

Although their use for regular passenger services disappeared, ocean liners took on another lease of life. Like horse riding and sailing, they have shifted from their basic utilitarian transport function to one of luxury leisure transport. The primary purpose of people who go cruising or yachting or horse riding is not to get from one place to another. They travel for the sake of travelling.

The epistemic shifts of the 1860s and 1960s

Paradigm shifts occur where the steep slope of the S-curve of adoption of an innovative technology crosses the matching steep slope of the reverse curve of a declining technology (see Figure 6.5). In 1835 sail was the norm paradigm for global trade. There were no regular steamship services. By 1890 steam was the norm paradigm and there were few regular international services offered by sail. Windjammers would continue to run down the Roaring Forties carrying wheat and wool from Australia and New Zealand to Europe until the Second World War, but that was because special conditions obtained. The winds were a strong and constant source of power for sail while the seas they built up put enormous stress on propellers because they regularly came out of the water. It was in the 1860s that there was a serious choice between steam and sail for international shipping services. Large ports had

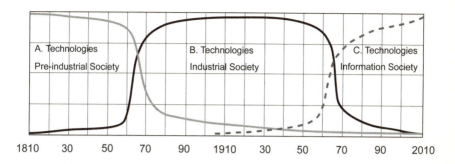

Figure 6.5 An idealized overview of paradigm shifts in transport from 1810–2010

A-type transport derives energy from wind and water and humans and animals, B-type transport is powered by steam derived from coal and C-type transport is fuelled by oil.

newspapers that carried notices of ship departures and arrivals and in the 1860s on international routes these were as likely to be steamers as sailing ships.

Comparison of Figures 6.4 and 6.5 shows paradigm shifts in transport that conformed and contributed to the epistemic shifts of the 1860s and 1960s. The key paradigm shifts were in the traffic and infrastructure networks enabling transport. Paradigm shifts in the regulatory, skills and auxiliary services networks necessarily followed. The communications industry underwent its own paradigm shifts in the period 1810–2010 and although the overall pattern was not quite the same, it profoundly impacted on the communications networks that enabled transport.

Until the mid-19th century, communications systems essentially were transport systems. If anyone wanted to communicate by speech over distance, they needed to travel in person and if they wanted to communicate by written or printed word they had to use postal or carrier systems. Communications paradigms were directly impacted by the paradigm shifts in transport and transport paradigms were in turn impacted by shifts in communications paradigms.

Interurban postal services shifted from using stage coaches to using trains and in consequence were able to extend overnight service over longer distances than was possible by stage coach. The postal service had its own paradigm shift with the introduction of the adhesive postage stamp in 1840. 'The Penny Black' as it was called meant that the sender of a letter prepaid one amount to have it delivered no matter where, within the British Isles. Simple as the idea was, it solved the age-old problem of who paid what for the delivery of mail and it was the first real revolution in the 4000-year-old history of postal systems. The efficiency that came with using stamps brought the price of mail down to a point where it was in reach of the mass of people at a time when railways were beginning to make mass transport possible. Postal services expanded with railway services and this brought further interaction. Mail trains were introduced with coaches that sorted the mail while it was in transit.

The first mail train ran in 1838 and the last one ran in 2003. At their peak from the 1860s to the 1960s, mail trains criss-crossed through the night between the cities of industrial societies much as the 'red eye' flights now cross oceans and continents. Besides mail and sleeping passengers, the mail trains carried national daily newspapers. The departure times of the mail trains were the deadlines for the national editions of newspapers so that they could be dropped off at each city en route in time to be sorted and delivered to homes, paper shops and news stands ready for breakfast and the morning rush hour.

Every town in 18th century Europe had hand presses that turned out broadsheets or local newspapers or small editions of books for the limited number of people who could read and afford them. In 1813 *The Times* began to use steam power for printing which, together with the growth of railway transport and mass production, dropped the price of newspapers to a point where most people could afford them. This occurred at a time when compulsory education as a public good was being established in industrial countries creating a 'mass' reading public.

People who travelled on the top of stagecoaches caught the wind and weather. Even those who travelled 'coach' had to suffer the sway and bounce of passage over rough roads. Trains provided a smooth, seated, windless journey where it was possible to pass the time reading. The railways created a reading public. They extended the mass medium of print from newspapers to magazines and cheap editions of books and started the entertainment services that passengers now expect.

In 1844, telegraph wires were strung alongside the railway lines between Washington, DC and Baltimore. The first message sent by its inventor Samuel Morse was 'What hath God wrought!' Well might he have asked. Morse had not only launched a way of sending messages over wires, he had also launched the meta-idea of modern telecommunications.

The telegraph was intended to control railway traffic and infra-structure and facilitate rail services. This it did most effectively and telegraph posts and wires became part of the railway infrastructure. However, it quickly became evident that the telegraph was of much wider use for transmitting urgent news. The telegraph grew at the rate we associate today with the internet. By 1866, a transatlantic cable was complete. Where news of events had depended on how fast a ship could cross the Atlantic, it now arrived virtually immediately. Communications had been uncoupled from transport and become global. By the end of the 1860s the continents of the world were linked by submarine cable and the telegraph meant that messages could be sent across the world in minutes.

Many communications technologies have, like transport technolo-gies, approximated to the S-curve of innovation in which the period of rapid adoption lasted about a decade. It happened to film in the 1920s, to radio in the 1930s, to television in the 1960s and to the internet in the decade between 1995 and 2005. However, paradigm shifts in communications technologies in the 20th century do not show the clustering around the 1960s that is evident in transport technologies. The telephone has been going from strength to strength for over 100 years. From its invention by Alexander Graham Bell in 1876 it has increased its reach and capability. From being a medium of urban communications it has become an international medium. Now, as the vehicle for the internet and as it acquires broadband capability, it is subsuming other communications media to become a vehicle not just for sound and text, but for image, radio, film, television, books, newspapers, magazines and virtual realities. It is linking with radio to become location free, a means of communicating globally from anywhere to anywhere simultaneously by sound, image and text.

From its start, telecommunications has been linked to transport. Radio in its many manifestations (radar, GPS equipment, radio navigation systems and radio beacons) remains the main medium for sea and air traffic control. In the growing use of telecommunications to direct, control and communicate with transport traffic, a shift in emphasis in the relationship between transport and communications is emerging. Whereas in the past improvements in transport led to improvements in communications, it is now improvements in telecommunications that improve transport.

The episteme of globalization

Lost tribes have been found in Brazil that had no concept of the world outside the forest they knew and travelled across and lived in. This was the space of their episteme, one that for them ended with their discovery and their realization that there were other people than themselves, other languages and customs than their own and that the world was much bigger than they had imagined. Something similar has been happening to the citizens of nation states as they realize they are in fact part of a global community, that no country stands alone, that no man or woman is an island and that everyone is subject to global warming.

The historical picture of epistemic change over the last two hundred years that is encapsulated in Figure 6.5 is from the perspective of developed countries. The epistemic changes of the 1860s and 1960s took place in countries such as the United States, Germany, the United Kingdom, Japan and France. They are referred to as industrialized or developed countries as distinct from the countries that have yet to complete an episteme as an industrial society and can, therefore, be referred to as industrializing or developing. But does a lost tribe have to go through an industrial revolution before it can be part of modern Brazil? Does it then have to learn to be part of Brazil before it can go global?

Paradigms and the NETS

The networks enabling transport are paradigmatic. A critical auxiliary services provider (ASP) that is networked to be available to all motorized transport systems is that which provides the fuel for the energy needed for transport to happen. It was the shifts in this paradigm that came with the use of coal instead of animals and wind in the 1860s and then with the use of oil instead of coal in the 1960s that triggered a domino effect in transport and communications paradigms that brought about the epistemic shifts of the industrial and information revolutions. Now problems are appearing in the use of oil. We are close to peak oil which means the amount of oil that is left is equal to the amount consumed. From now on, the more oil-powered transport continues to expand, the more expensive it will become. It is also becoming apparent that the switch to carbon-based fuels that began a century and a half ago is provoking climatic changes that will massively damage the environment transport serves. The continued growth of transport as we know it will be destructive of transport as we know it. Such a paradox precedes a paradigm shift. The next chapter considers some of the alternatives.

7

Intelligent transport: the new communications technologies

Nanoscience is the study of phenomena and manipulation of materials at atomic, molecular and macromolecular scales, where properties of matter differ significantly from those at a larger scale. Nanotechnologies are the design, characterization, production and application of structures, devices and systems by controlling shape and size at nanometre scale. (The Royal Society, 2004)

Nanotechnology can make big things as well as small things. An attractive approach is to use convergent assembly, which can rapidly make products whose size is measured in meters starting from building blocks whose size is measured in nanometers. It is based on the idea that smaller parts can be assembled into larger parts; larger parts can be assembled into still larger parts, and so forth. This process can be systematically repeated in a hierarchical fashion, creating an architecture able to span the size range from the molecular to the macroscopic. (Merkle, 1997)

Today HyperReality is at the stage television was at when John Logie Baird was experimenting with it: a technological capability waiting to be developed but still in the laboratory. It is a medium for communication between the real and the virtual, between human and artificial intelligence and between fact and fiction. (Tiffin in Tiffin and Terashima, 2001)

Introduction

Chapter 6 looked back 200 years at the relationship between transport, communications and globalization. This chapter looks to the future. Where Chapter 6 plotted the steep curves of rapid adoption of new transport technologies, this chapter looks at the initial shallow slopes of early adoption that follow invention in a search for the technological trends that could accelerate to become the new transport paradigms of the future. The steam engine was invented in 1712 almost a century and a half before it caused a paradigm shift in the application of energy. The likelihood is that the transport technologies of the future have already been invented, but the history of transport is full of marvellous inventions that never made it to the steep slope of adoption.

In 1937, exactly one year after a trans-Atlantic Zeppelin service had been successfully inaugurated, the *Hindenberg* burst into flames as it was docking. The occasion was marked by one of the first live external radio broadcasts, so that the disaster was transmitted in vivid and shocked terms nationwide. The impact was similar to that of the events of 11 September, 2001. Until this disaster, the future for aviation appeared to be with dirigibles, but the negative impact of the radio broadcast meant that Zeppelins have never been taken seriously since.

There is a widespread assumption in the transport industry that demand for transport of goods and people will continue to grow as it has done since the Second World War. The response to this at the global level is to build more ships and aeroplanes and make them bigger and faster. At the national and urban levels it is to make more motor vehicles and roads.

Since the 1860s the energy for transport has come from burning fossil fuels. The weight of scientific opinion is that this is causing climatic change, which if unchecked could trigger irreversible global warming (Kolbert, 2006). But we seem to be locked into epistemic thinking of a world run on oil. Like the sorcerer's apprentice, we seek solutions by doing more of what we are already doing, which was causing the problem in the first place. Transport needs to look for solutions in new technologies, but what are they?

Nanotechnology

One technology that could prove as revolutionary as the steam engine is nanotechnology. This is the technology of the extremely small, where objects are measured in nanometres. Drexler popularized the idea of nanotechnology when he envisioned what the world could be like if we made things directly from atoms and molecules (Drexler, 1986). Instead of large scale manipulation of the physical environment with huge machines, nanotechnology allows people to fabricate what they want with micro-machines. It becomes possible to build by the accretion of atoms rather than from the reduction of their raw mass. What is manufactured could be made, where it is wanted, with nanofactories.

How viable is such a technology? Following Drexler there have been numerous attempts to design a nanotechnology manufacturing system (Bishop, 1996; Merkle, 1997; Hall, 1999; Phoenix, 2003). A nanofactory can be conceptualized as something the size of a desktop printer linked as a peripheral to a PC. To make a pair of gloves you might log onto the internet and search for a suitable glove design, select a shade of colour and have your hands measured for a perfect fit. Then you purchase and download the design. Alongside the Save and Print commands on your PC there could be a Fabricate command. You select it and the nanofactory sitting next to the printer produces the gloves. The design, not the labour, not the raw materials and certainly not the transport, would be the principle cost of manufacturing the gloves (Phoenix, 2003). Unless some exotic chemicals were involved, the bulk of raw materials (in nanotechnology known as feedstock) could be derived from the elements of carbon and hydrogen that are all around us. The nanofactory processes molecules from a feedstock of elemental raw material. From these molecules are fabricated nanoparts which are joined to form nanoblocks. These basic building blocks join together in progressively larger units to form the finished product. Hall (1999) outlines how the nanoblock units could be self-replicating to 'grow' the final product.

Phoenix (2003) calculates that it would take about an hour to make something the size of a torch or a kettle. He also argues that things manufactured by nanotechnology would be stronger, have less volume and mass, have more components and complexity, be easier to use and need less energy than comparable products manufactured by conventional factory technology. This would be by such an order that conventional manufacturing would be unable to compete.

A desktop nanofactory could duplicate itself and double its size to fabricate larger objects and double its size again and again. With convergent assembly, nanotechnology could be used in the manufacture of relatively rigid structures made from rigid components such as computers, cars, aeroplanes, bicycles, buildings and furniture (Merkle, 1997).

Forests are ripped apart for pulp for paper, props for mines and planks for joinery. Vast areas of land are quarried and mined. The resulting raw materials are brought together from different parts of the world in factories where they are reduced, refined, assembled and reassembled. Then the finished products are distributed back to the markets of the world. This is how industrial society functions. It is made possible by bulk carriers and container ships. If the promise of nanotechnology is realized, it could mean a return to an era of cottage industry where most of a society's needs are found locally. Point-of-use manufacturing could eliminate supply chains. Bulk transport of raw materials and manufactured goods would decline. Products coated with nanofabricated substance that converts light to power would reduce the use of fossil fuels. The need for power grids would decline.

Clever clothes

Clever clothes refers to the application of IT and nanotechnology to what we wear (Tiffin and Rajasingham, 1995; Tiffin, 2001; Tiffin and Terashima, 2001). This becomes increasingly feasible as cellular radio environments become ubiquitous and computers continue to shrink. The trends are evident in mobile phones, the growth of computer games and interest in the idea of wearable computing.

Mobile networks and Bluetooth networks are already creating the radio infrastructure that clever clothes need. Cellular radio, like nanotechnology, is essentially fractal in nature. Its levels extend from the footprints of satellites down to the Bluetooth environment in which neighbouring computers talk to each other.

Wearable computing today means carrying PCs as accessories in the manner of a handbag or wrist watch or a mobile phone. Sometimes linked to a head-mounted video unit, wearable computers can be carried in a backpack, clipped to a belt, or worn on the wrist. From their first appearance in the early 1960s, wearable computers have found applications in transport in such areas as vehicle inspection and aircraft maintenance (Amon *et al*, 1997). In the lead in wearable computing is mobile telephony. In its third generation this technology is also an internet-linked computer, able to transceive video, audio and text.

The idea of clever clothes refers to what will be possible when nano-computers and nanotransceivers are woven into the fabric of what we wear. Drexler (1986) imagined a form-fitting costume like that worn by spiderman, which completely enveloped its wearer. Clothes have a utilitarian purpose to protect and a communications function to say something about the person wearing them. Clever clothes would enhance both these functions and introduce some new ones. A basic nanosuit would be lighter, less bulky and stronger than conventional clothes. It would be customized to fit with the precision of a mould, but padded, ribbed and textured on the outside to give the wearer the kind of figure and persona they wanted to project, whether it be an Adonis or a Venus, Batman or Catwoman, or themselves, only younger, more attractive and with ethnic and sexual characteristics that were chosen rather than given. The clothes would have a layer of artificial muscle woven from diamonoid fibres able to push as well as pull, but with far greater power than that of humans. Powered by nanoelectric motors and controlled by nanocomputers, they would stretch, contract and bend and so greatly increase their wearer's ability to lift, carry and manipulate objects. Labouring practices on building sites or in heavy duty situations in a dockyard or on a ship in bad weather would be strengthened. Sports that depended on running, jumping and swimming would acquire new levels of skill.

Clever clothes made of nanocomputers woven together with nano-fibres would provide a parallel processing network capable of support-ing AI that could coordinate its movements to match those of its wearer. Unlike conventional clothes that are supported and held in place by the people who wear them, clever clothes could support and amplify the movements of the wearer. It becomes possible to imagine a future where people can walk further, run faster, jump higher, and hit a backhand with extraordinary force and accuracy. Public concern would switch from what drugs athletes were taking to what clothes they were wearing. The elderly and disabled would have renewed mobility. People could roam over rugged terrain, climb high mountains and swim in rough seas with impunity. They might even be able to realize one of humanity's oldest fantasies and fly like birds. With nanotechnology it should be possible to devise winglike extensions to the arms and the suit would have the musculature and intelligence to imitate the movements of bird flight.

Drexler (1986) saw nanosuits as having their own nanopower plant which would be little more than a small backpack on the suit. They could also draw on solar energy. DaimlerChrysler are researching spray-on nano solar paint that can absorb sunlight to release electrons that augment power. They would also make vehicles self-cleaning, scratch-proof and able to change colour. Such properties could be applied to clever clothes. They would always appear new and they would be energized by batteries supplemented by ambient light so that they could function autonomously. When people take their clothes off, they hang them in a wardrobe as so much lifeless matter, but clever clothes would be able to stand and move by themselves and act robotically independently of any wearer. People would be able to step out of their clever clothes and leave them to perform tasks by themselves or in a team of like-minded smart suits. Looking at a suit of clever clothes in action it would not be possible to tell if there were anyone at home inside, or if there were, who it was.

Nanotechnology can replicate a product endlessly. It would be possible to duplicate a smart suit. A uniform of clever clothes made for one soldier could be duplicated to become a regiment that would behave exactly as one soldier. The film 'Matrix' uses the idea in fight scenes, where the hero is endlessly faced with ever increasing numbers of humanoid robots who appear as clones of the villain. Such uninhabited suits need not be limited to human dimensions. A giant-size copy of an emergency worker could stand over a railway crash to lift carriages.

The surface of the suits would contain nanotransceivers able to tune in to different parts of the electromagnetic spectrum. Theoretically it would be possible to receive reflected light on one side of the suit and transmit it from the other side at a targeted person so that to them the suit would seem invisible. The suit could augment or diminish the impact of external stimuli on its wearer. Extremes of heat and cold and violent blows on the exterior of the suit could be tempered and it would be possible to see, hear and touch with great acuity. The suit would also be able to transceive signals between its occupant and the telecommunications environment. This means clever clothes could facilitate entry into HyperReality.

HyperReality

Nobuyoshi Terashima, when he was President of Japan's Advanced Telecommunications Research Laboratories from 1991–1996, led a team that developed a prototype of a virtual teleconferencing system that allowed people in different places to come together in a virtual space to do something. Terashima realized that this was a new kind of medium, one where real people in real space, with real objects and tools could interact with virtual people in virtual space with virtual objects and virtual tools. He called this 'telesensation' (Terashima, 1993) and used the term 'coaction field' to refer to the space where virtual reality and physical reality intersect to allow interaction. A similar concept, called 'teleimmersion', has since been introduced by a group of US researchers led by Jaron Lanier. Meanwhile Terashima had moved on to realize that the components of a virtual reality do not have to be extensions of humans or even of human imagination. They could be generated by AI and be artificially intelligent. He called the advanced form of reality that results HyperReality. This is how he describes it:

> The concept of HyperReality (HR), like the concepts of nano-technology, cloning and artificial intelligence, is in principle very simple. It is nothing more than the technological capability to intermix virtual reality (VR) with physical reality (PR) and artificial intelligence (AI) with human intelligence (HI) in a way that appears seamless and allows interaction.
>
> The interaction of PR and VR in HR is made possible by the fact that, using computers and telecommunications, 2D images from one place can be reproduced in 3D virtual reality at another place. The 3D images can then be part of a physically real setting in such a way that physically real things can interact synchronously with virtually real things. It allows people not present at an actual activity to observe and engage in the activity as though they were actually present. The technology will offer the experience of being in a place without having to physically go there. Real and unreal objects will be placed in the same 'space' to create an environment called a HyperWorld (HW). Here, imaginary, real and artificial life forms and imaginary, real and artificial objects and settings can come together from different locations via information superhighways, in a common plane of activity called a coaction field (CF), where real and virtual life forms can work and interact together.
>
> Communication in a CF will be by words and gestures and, in time, by touch and body actions. What holds a coaction field together is the domain knowledge (DK) which is available to participants to carry out a common task in the field. The

> construction of infrastructure systems based on this new con-
> cept means that people will find themselves living in a new
> kind of environment and experiencing the world in a new way.
> (Terashima, 2001)

The clever clothes concept of intelligent nanofabricated body condoms is compatible with the concept of HyperReality. A person could partici-pate in a coaction field in clever clothes in person, or as a telepresence or as a robotized automated version of themselves. It would be possible to have multiple presences in a coaction field. Pilots on duty at a port could perform their traditional duties guiding ships in and out by having a telepresence on a ship's bridge. In effect this would mean occupying a nanosuit as a telepresence on a bridge to sense the state of things from the perspective of a ship's captain, while in fact the pilot could be standing in a room in a port authority. Given a tricky docking situation involving several tugs, pilots could have telepresences on board each of the tugs so that they could switch viewpoints.

The telephone is a primitive form of HyperReality that allows people in physical reality to talk to virtual voices. HyperReality could be thought of as telephoning a three-dimensional version of yourself, along with the place you are in and the things you are handling and it allows you to touch and talk and interact with whoever you are phoning and with the place they are in and the things they are handling. There is a critical proviso. You can only interact within the specific domain of knowledge of a particular coaction field. So, for example, a patient and doctor in separate places could interact in a medical coaction field. To the doctor in a surgery the patient and the bed and room they are in are virtual, to the patient in their own bedroom the doctor in the surgery is virtual. However, they can only interact with each other in terms of medical conditions. They could not, for example, cook a meal together. Similarly a port pilot and ship's captain could only interact in connection with steering a ship in a particular stretch of water. An Automobile Association emergency operator could only interact with a stranded motorist to look at their car and advise them what to do. If the car driver sitting in a car and the emergency operator sitting in an office are both wearing nanosuits, the emergency operator would be able to see what the driver sees as they look through the lenses that are the eyes of the suit. If the driver's suit was slaved to the operative's suit then the operative could carry out repairs from their office.

Clever clothes make it possible to drive a vehicle such as a van or an aeroplane or a ship without physically being in the vehicle. Aeroplanes and ships now make extensive use of automatic pilots. Military forces use drogues. The principal of using AI to pilot can be extended so that vehicles can perform complete journeys without humans doing any piloting. When things go wrong in transport it is usually because of human error. HyperReality means that a human (or an AI) in a central control point can control a vehicle in telepresence. At the scene of a crash, emergency workers in clever clothes could call on the help of doctors or engineers who were not actually at the scene to look at a situation. They could look at the problem through the 'eyes' of the emergency workers and advise them what to do. To some degree, a distant doctor wearing a similar suit to an emergency worker would be able to take over at the site to manipulate an injured person or check his or her condition.

Tourists could make virtual visits to dangerous places without risk of disease or kidnap. Doctors could make virtual rounds of their patients. People could play virtual sports and attend a virtual Wimbledon. The infirm could pay virtual visits from their hospital beds. The elderly and infirm could travel with ease supported by their nanosuit, their blood pressure endlessly monitored and emergency services could be called on by their clever clothes if they became ill.

Artificial intelligence (AI)

There is no unambiguous definition of what intelligence is. We understand it by our own lights as making sense of things and being able to understand, perceive, remember, learn, reason and solve problems. Intelligence is also in the way we talk, listen, read and write. We have tried to replicate these competencies in AI, but they have proven more complex than we ever imagined. The goal epitomized in the story of Dr Frankenstein to make an intelligent creature in our own image eludes us, but remains the principle vision behind the drive towards 'true' or 'strong' AI.

Joseph Weizenbaum is quoted as describing the ultimate goal of strong AI as being nothing less than to build a machine on the model of man, a robot that is to have its childhood, to learn language as a child does, to gain its knowledge of the world by sensing the world through its own organs, and ultimately to contemplate the whole domain of human thought. Strong AI catches the imagination, but by no means all AI researchers view it as worth pursuing. Some critics doubt whether research in the next few decades could produce a system with the overall intellectual ability of an ant.

If we stop trying to compare human intelligence in all its richness with AI and focus instead on some of the specific activities that humans and computers can do, which are held to be aspects of intelligence, such as remembering things, doing sums and playing chess, then computers are already more intelligent than we are. In what is called 'applied AI' or 'advanced information-processing' we have already developed a symbiotic relationship with computers not unlike that which we have with domestic animals.

A part of AI that has been widely adopted for commercial use is that of expert systems (Giarratano and Riley, 1998; Jackson, 1999). An expert system advises on problems in some specialized field. They learn from their mistakes and from capturing knowledge from not one, but many experts.

Theoretically if all the experts in some particular field of expertise in some particular organization interacted with the same expert system, that expert system would know as much as all of them put together and more, because it would rationalize the sum of knowledge and begin to generate new expertise. In restricted fields of expertise, expert systems come to outperform human experts. The application of expert systems to specific worlds of expertise is essentially the same as the application of HyperReality domains of knowledge in coaction fields.

We use expert systems in transport like medieval hunters used animals: horses to gallop them to the field, dogs to sniff out and start up the game and falcons to make the kill. These animals all had skills superior to humans in specific fields, but it was the human hunter who had the overall purpose and vision of the hunt and coordinated what happened. So in air transport it is possible for expert systems to schedule flights, organize cargo placement and maintenance and to pilot aeroplanes. However, the overall management of an airline, the sense of purpose and the coordination of different expertise remains with humans, but for how much longer?

Having a human in charge allows for the unexpected, but expert systems learn. In time they accumulate more experience of the unexpected than humans and so become more reliable in reaction. They know and follow all the rules and regulations. So will there come a day when transport vehicles are entirely driven by AI robots and humans, if they are physically present at all, are only there as passengers? The principal obstacle to such development would seem to be the extent to which communications between traffic, and between traffic and infrastructure, is in some medium that is perceptible to humans but not to IT.

Traffic on a modern road system is driven by many human intelligences that selfishly compete with each other to optimize the use of the road for the individual benefit of each vehicle. The result is an inefficient system prone to accidents. A road infrastructure could be managed with AI that had the objective of maximizing the benefits of the system for all the traffic using it. Traffic control AI would link to road infrastructure AI to control the interval between vehicles and how they switched lanes and got on and off the system across the whole network. The overall intelligence of the system would have the computing capability to perceive from digital data where there were traffic build-ups and to adjust flows across the entire system to draw them down before they became critical. There would be no need for cars to drive with sufficient distance between them for the driver to react. Cars would be able to drive with only a few centimetres between them because the system would know what was happening across the whole network and the same intelligence would be driving all the vehicles.

New technologies and epistemes

How likely are the technologies described above to eventuate and affect the field of transport? These technologies are extensions of trends that are already set and the environment that makes them possible is coming into place. Computers continue to become more powerful and are increasingly interlinked. Telecommunications become broadband, mobile and ubiquitous. Cameras and sensors linked by telecommunications to computers grow like mushrooms in the atmosphere of suspicion that has enveloped the world since 11 September 2001. RFID tags that say what or who something or someone is, where they are and what their purposes and history are, are attached to more and more people and things (Poirier, 2006). The growing conjunction of RFID, sensors and observation cameras linked by telecommunications to computer hubs makes it possible to quantify and image the status of traffic, cargo and passengers at a global level so that interventions can be made at every level of a supply chain. It is a massive advance on the way only half a century ago transport depended on people trying to tally what they could see. Such systems can also record and accumulate permanent records of everything that is transported. This makes it possible to study where systems can be improved. It also means that it becomes increasingly difficult for an individual to have a private life. In the wrong hands it could create an Orwellian society.

At some point in a future that may not be so far away we shall find ourselves in a new episteme. It is hoped that in the process we will have acquired intelligent transport systems that allow safe, sensible sustainable development of the planet for humans as a whole. The next four chapters look at some of the problems in the way of this at different levels of different modes of transport and at some of the ways these might be circumvented.

Of course, if the dreams of strong AI are realized, it could be that the episteme of globalization will be the worldview of artificial life (AL), where transport exists for the benefit of IT systems and humans are the buttons that computers press and the things that can be blamed when systems go down.

8

Seeking space: water and air transport

Primus circumdedisti me. (Authorized on the coat of arms of Juan Sebastián de Elcano by the King of Spain to commemorate the first circumnavigation of the globe. It translates: 'the first to encircle me.')

Space, the final frontier. These are the voyages of the Starship Enterprise on its five-year mission to explore strange new worlds, to seek out new life and new civilizations, to boldly go where no man has gone before... (Opening lines of the 'Star Trek' TV series)

We wait in nondescript boarding lounges, walk down metal corridors and lever ourselves into the narrow seats of a small cinema where we watch Hollywood films on low definition screens while unsmiling staff push trays on our laps bearing an assortment of inedible food that we are not expected to eat... After a few hours we leave the cinema and make our way through another steel tunnel into an identical airport in the suburbs of a more or less identical city. We may have flown thousands of miles but none of us has seen the outside of the aircraft and could not say if it had two, three or four engines. All this is called air travel. (J G Ballard in *The Guardian Weekly* (London), 3–9 June 2005)

Introduction

On 8 September 1522 the *Victoria* limped into harbour having completed the first circumnavigation of the world. She was the last of the five ships that had departed three years before with Magellan. He had been killed in the Philippines and it was Juan Sebastián de Elcano, whose motto is cited above, who led the 17 crew that were all that was left of the 250 who had set out. They had departed in one direction and returned from the other and left no doubt that the earth was a globe. So globalization began with water-borne transport.

Almost 500 years after Magellan, Bertrand Piccart received the Magellan Award from New York's Circumnavigation Club for the first non-stop flight around the world by air. Piccart did the journey in 20 days in a balloon. Like Magellan he used the wind. The skyways and seaways are nature-given and, for most of history, ships have used the energy in wind and water free of charge. It is only in the last two centuries that vehicles have been propelled through the air and across land and sea with energy generated from fossil fuels.

From its beginnings, global water transport moved people as well as freight. By the time the slave trade was abolished in the British Empire in 1807, some 15 million slaves had been shipped from Africa to the Americas. This was followed by yet another vast seaborne migration as the poor of Europe sought a new life in the Americas and Australasia. People still move around the world in their millions, but now most of them do it for tourism and they go by air. The main purpose of shipping has become freight not people. The 'poor and wretched' are still there and still 'yearning to be free' but their migrations are restricted and tend to end in refugee camps (Castles, 2003).

The first attempts at manned flight took place on 21 November 1783, when Pilâtre and d'Arlandes flew over Paris in a Montgolfier balloon. Balloons were used by Napoleon and in 1852 Henri Giffard of France flew an airship driven by a propeller powered by a steam engine at a speed of 10 kilometres per hour over a distance of 30 kilometres. The *Hindenberg* explosion in 1937 marked 150 years of experience in using balloons for flight. Ballooning grows in popularity as a sport and airships are still in use and deployed in such activities as selective logging to avoid environmental damage in forestry operations. However, even with inert helium to replace hydrogen as the lifting gas, confidence in this technology has not yet returned and the way of aviation has been with the kind of controlled powered flight demonstrated in December 1903 by the Wright brothers in their biplane *KittyHawk*.

Until the large jets created a plateau for aeroplane technology in the sixties the story of aviation was one of experimentation in aeroplane design in the two world wars and exploration of possibilities in peacetime. Adventurous people sought to be the first to fly between two cities, then across seas and oceans and continents (Wohl, 2005). As with sea transport, trade followed exploration. Many famous airlines such as Lufthansa, British Airways and Qantas were established in the early days of flight as they sought to create services based on the routes that the pioneers opened up. Early international aviation was based on reciprocity between countries as routes were established between them. Strict bilateral agreements allowed them to control and share the available business. As governments embraced free market philosophies in the late 1990s, liberal bilateral and plurilateral agreements were introduced. In the fierce competition that ensued, many of the major airlines went bankrupt.

What is the future for air and water transport? What is the next paradigm shift in these transport modalities likely to be? This chapter examines the NETS of water and air transport at their different fractal levels and then looks at the changes that IT could bring about.

The fractal levels of water transport networks

At the global level the oceans and continents impose an infrastructural network on water transport. The oceans are nodes and the links are the channels between them. The history of maritime discovery is one of finding ways to get from one ocean to another to open trade between them. The routes around the Cape of Good Hope and Cape Horn meant making long diversions and facing extreme weather conditions. The Suez and Panama canals were built to resolve these problems. Unlike the cape routes the canal routes favoured steamships and triggered the paradigm shift from sail to steam which revolutionized global transport networks.

Shifting down a fractal level to the regional level, the oceans become networks. Links at this level are the shipping lanes that cross the oceans and nodes are the riparian trading countries. Also at this level are continental seas such as the Baltic, the Mediterranean and the China Sea and the continental river and canal systems such as those of Europe and the St Lawrence seaway in Canada.

At the national level, countries have networks of coastal and inland waterways that link ports and towns. Vessels are normally smaller at

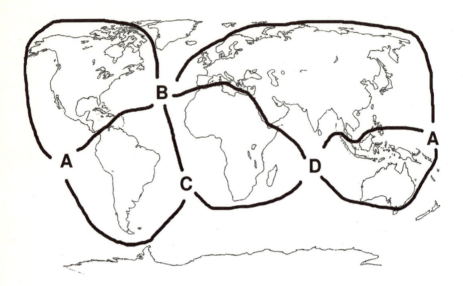

Figure 8.1 The global level network of maritime transport

A is the Pacific, B the North Atlantic, C the South Atlantic and D the Indian Ocean. The northern AB links across the north of Canada are the legendary North West Passage and the BA link to the north of Russia is the North East Passage They were only available in good summers, but global warming is increasing their use. The southern AC link south of South America is the passage round Cape Horn. The southern CD link around Africa is the Cape of Good Hope route and the southern DA link south of Australia is the passage through the Bass Straits. All are notorious for storms. It is the middle links that take the most traffic. These are the Panama canal (AB), the Suez canal (BD) and the Straits of Malacca (DA). The BC link is the broad passage between Cape Palmas (West Africa) and Cape São Roque (Brazil) that links the North and South Atlantic.

this level to access shallow draught ports, small harbours and narrow canals. Accessing national from international networks may, therefore, involve reshipping. Archipelagic countries like Indonesia and the Philippines depend on networks of ferry services to link the many islands in their territories.

The next fractal level of water transport is found in the canals, creeks and inlets of such cities as Amsterdam, Bangkok and Venice. Yet another shift in boat size may be needed for these waters. The people who work the boats at this level often live on them. Boat people, along

Figure 8.2 The site level: water transport in an apple packing factory

The apples are moved from one processing function to another by water, which is not only gentler than human handling but a means of washing and disinfecting the apples. Photograph courtesy of R Palmer.

with others who live on or close to water, are increasingly at risk from the rising sea levels and storm surges that come with global warming.

The fractal levels of air transport networks

Jet aeroplanes ascend and descend through the troposphere where most weather occurs, but cruise in the stratosphere where air density is low so that less fuel is needed and conditions are relatively homogeneous and stable. With the advent of the Boeing 777ER and the Airbus A380 it becomes possible to fly half way round the world non-stop and to follow a great circle route directly between any two airports. The things that cause flights to deviate, such as severe storms, major volcanic eruptions, wars or the refusal of a country to allow overflights, are essentially temporary in nature, unlike the land masses that block global transport by water. It is possible, therefore, to think of global air transport as a potential level playing field for the forces of globalization where all airlines operate in a relatively uniform space to which all air traffic has physically equal access and any two airports' nodes can be linked.

As with water transport, there are different fractal levels of air transport each of which involves a change of gauge. First there is the level of international flights that corresponds with the global and regional levels of water transport. Here the nodes are nations represented by their international airports where immigration and customs have to be cleared to allow entry to a country and access to its national transport systems. To continue travelling within a country by air a traveller shifts down a fractal level to that of domestic air transport networks which are equivalent to the national levels of water transport. This means either transferring from international airports to domestic airports or from the international wing to the domestic wing of an airport. International air networks link nations. Domestic air services link the cities and towns within a nation. The size of domestic airports and the planes and runways used is normally on a different scale to that of the international level. There is also a third fractal level in aviation in the small planes and helicopters that land on airstrips and helipads.

At the international level the trend is towards bigger more fuel-efficient planes carrying ever bigger payloads of people or freight over long distances. Where free trade has made for competition between airlines, the cost of flying between international airports has become cheaper per passenger kilometre than for any other form of transport. Air fares at the national level compete with road and rail fares for passengers. The most expensive of all modes of transport is that of the small plane. There is no aviation equivalent to the motor car in terms of cost, ease of use and the ability to travel from door to door in the same vehicle.

Changing aircraft gauge is inefficient and, for passengers, inconvenient and tiresome. It means unloading passengers and baggage from one plane at one gate and loading them on another plane at another gate, quite likely at opposite ends of an airport if not in a different airport. The process has become more tedious with increased airport security. It is possible to go through up to eight security checks when changing flights. An international traveller going to a medium-sized town or city that has no international airport may find themselves over-flying it as their international flight takes them to an international airport where they then change to a domestic flight that flies back over the ground they have just covered. Shifting between the fractal levels of air transport extends the time and distance involved in a journey by air and the costs of flying increase.

Long-haul travel by air from one side of the globe to the other has reduced in cost relative to average wages in developed economies and so allowed growth in global air traffic. Where departures and destinations are located along major air routes serviced by a multiplicity of competing carriers, the cost of air travel is low. However, when a passenger deviates from routes with high frequencies, switches from one carrier to another, and downsizes in aircraft type, fare charges per kilometre travelled increase. When the fare collected for a multi-sector air ticket is divided between the various carriers on a distance-carried basis (pro rata), the amount allocated to smaller carriers flying short-haul connections to the major carriers flying between hub airports can be insufficient to cover their operating costs let alone make a profit.

Infrastructure networks in water and air transport

The infrastructures of NETS of air and water transport is primarily in the nodes. Ports, docks, quays, slipways and moorings are the fixed nodes of water transport infrastructures where water traffic begins and ends, freight and passengers are loaded or unloaded, vessels refuel or undertake maintenance and cargo and passengers transfer to different shipping gauges or other modes of transport. Landing strips, runways and airports, where people and freight are drawn together and processed for embarkation or disembarkation and planes are serviced and refuelled, are the nodal infrastructures of air transport. Ports grow up around docking facilities. Airports grow around landing facilities. Over time the increasing size of ocean-going vessels has meant an increase in the size of such facilities and this has forced ports to migrate downstream or build out to sea. Similarly the growing size of aeroplanes, the distance they need in which to land and take off, the frequency of scheduled flights and the noise they make has meant an increase in the size of airports and a move away from the built-up urban areas they serve. Several major airport developments have involved new sites dredged from the ocean to form artificial islands adjacent to major population centres and connected to those centres by high speed surface transport links.

Figure 8.3 A Panama canal lock

Marine architects take into consideration the physical limits of the Panama Canal when designing vessels. Those that fit are known as Panamax vessels. Larger vessels ply the Cape routes on round-the-world services taking advantage of economies of scale, especially in the high-value liquid bulk and container trades. Photograph with the permission of GFDL.

The seas and oceans and the atmosphere are there for the using and unlike land transport need little in the way of linking infrastructures to channel traffic. Ships and planes navigate along notional lanes and paths and are guided in this by communications infrastructures. Traditionally the communications nodes have been lighthouses, marker buoys, beacons, wind drogues and landing lights, but increasingly they become radar beacons, radio control stations and GPS satellites.

There are physical infrastructural links for water transport at the national and urban levels in the form of canals and dredged channels and at the global level in the Pacific and Suez canals. The latter still dominate the pattern of global networks of water transport. From time to time consideration has been given to moving the Panama Canal to the north so that it can become a sea-level canal. This would require massive earthworks and possibly nuclear detonations which would surround the scheme with environmental and political pressures. Alternatives are to increase the capacity of ships capable of using the Cape routes and improve land-bridge operations. Railways that span continents can link sea segments in global supply chains as between Europe and Asia, the East and West coasts of North America, and in Australia between Adelaide and Darwin. At a regional level canals and canalized river infrastructures allow water traffic to penetrate the heart of continents. The St Lawrence river and seaway system connects the Atlantic Ocean to the Great Lakes in the United States and links to the Mississippi navigation system. The Rhine river in Europe links Switzerland to the North Sea.

Seaports are located because of the natural advantages that come with high tides, depth of water, clear approaches and access to hinterlands. Airports similarly are located where there are natural advantages of flat land and benign weather. What, however, is critical to both air and sea ports is that they are adjacent to land transport hubs. In recent times it has also become critical to have good telecommunications facilities. Places like Singapore, Rotterdam, New York and Hong Kong are not only seaports and airports and terrestrial transport hubs, they are also teleports. They are multipurpose multimodal superhubs for transport and communications where each activity reinforces the other. Thus they become the headquarters of companies involved in transport, travel and trade and the banking, finance and insurance services that support them.

Ports need to be operational around the clock and this raises social and cultural communication issues. Good relations are needed with the communities around dockyards and airports so that they accept noise and movement and the presence of people of many nationalities and do not seek through local authorities to impose curfews on night operations. A just-in-time environment calls for flexible deployment of labour. The casualization of dockyard and airport labour in recent years has given rise to disputes as unions seek permanence of employment and access to lucrative overtime rates of pay, while management seek to retain a permanent core of competent skilled labour bolstered by on-call part-time labour. Such situations call for the high level negotiation skills that are part of organizational communications.

Good human communications and the cybernetic control that comes with IT improve the efficiency with which ports operate, but knowledge of traffic movements and good community relations, while improving efficiency, are not critical. Ports have been able to function, albeit less efficiently, in the days when the first news of the arrival of a ship was the firing of a signal cannon and planes announced their arrival by buzzing the control tower. What was then, and still is, critical information is what is on board the incoming ship or plane: for example enemy soldiers, avian flu, a bomb. Electronic documentation of cargoes provides advance notice of what a ship is supposed to be carrying, but this still needs verification. This issue faces airports as well as seaports. The level of verification applied is in direct proportion to the security assurance that can be given by the country of origin. Ports have to be secure environments. Assemblers of cargo have to be security vetted.

The pressure is on international hub ports to use IT to effect a quick turnaround of traffic. Airports are primarily concerned with moving people and ports with moving cargoes, but in both cases time spent quayside or at airbridges costs money. Turnaround, even for ships, is now measured in hours with the use of sophisticated handling equipment and smart information systems.

At the lower fractal levels of water and air transport, the pressures on infrastructure diminish. Canals, small ports and landing strips are available on demand. Small airstrips operate without airport control tower assistance. It is left to pilots to check on local traffic movements and to announce their intentions to each other. If they are flying visual flight rules (VFR) they must keep away from airspace that is used by aircraft flying under air traffic control. When approaching a controlled airport, they must come under the guidance of air traffic control. Similarly the crews of small boats operate lock gates and look after their own mooring and anchoring needs except when they are operating in a large port and come under a port control authority that may designate specific mooring areas and the navigation channels they must use.

Traffic networks

A traffic network consists of the floating nodes of vessels or aircraft that are operational in a particular water basin or airspace and thereby have a potential relationship with each other as well as with the infrastructure. Network links in water and air traffic are the communications that transmit information about the position, movements and intentions of ships and planes. In land transport the road or rail of an infrastructural link stretching out in front of a vehicle clearly demarcates the field of activity for traffic. At sea or in the air, however, there is no immediate physical infrastructure. Prior to the Second World War, when they were out of sight of land, airplanes and ships plotted their positions and courses with the aid of chronometers and sextants. Light was the primary signalling medium for the communications that linked ships and planes with each other and with the infrastructure. Traffic networks were what existed within line of sight of a pilot or navigator. When it was dark or foggy, ships and planes had problems finding their way and avoiding collisions or, in wartime, finding the enemy. War provided the impetus for developing radar (radio direction and range). This paints a picture on a monitor of the traffic network around a particular vessel that extends over the horizon beyond human vision and is largely independent of weather and daylight. Radar became standard navigational equipment from 1950. Now GPS devices enable precise navigation in three-dimensional space.

Ensuring that safe separation is maintained between airborne aircraft and between airborne aircraft and aircraft on the ground that are marshalled for take-off or taxiing to parking positions determines the limits to the flow of traffic on a particular route and in consequence the timetabling of flights. Until the late 1990s, radio communications used several frequencies for this, but on long-haul flights, atmospheric conditions could impair voice contact. Air navigation required large vertical and horizontal separation between aircraft on predetermined flight paths. It is now possible to use global mobile satellite communications (GMSC) that makes it possible for aircraft to fly closer together with higher levels of safety. In conjunction with radio directional beacons, GMSC can provide a line of guidance and the ability to home in on a target. GMSC also allows planes to deviate from flight plans if circumstances warrant. Rather than buffet headwinds and incur higher fuel burn, routes can be optimized for maximum use of tail winds and altitude can be adjusted to allow for the changing weight of an aircraft as it consumes fuel. GMSC also enables connections between moving objects such as ships, vehicles and aircraft through the interconnection of communications satellites, ground earth stations and landline telecommunications. The view of traffic networks expands to encompass all traffic.

Ships and planes do not always travel at similar speeds nose to tail across the oceans. They fall into line as they near a port or leave it and this is where transport congestion by air and water builds up. Time spent waiting to dock or gate is lost time for ships and planes and the

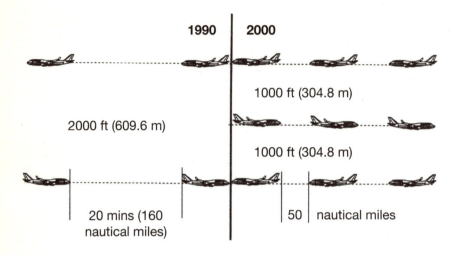

Figure 8.4 Increased use of airspace with GMSC

This shows the systems used to control aircraft movements on the ground and in the air around airports as well as on long-haul sectors. Diagram after Airways Corporation, New Zealand.

time docking facilities lie idle is lost time for dockyard and airport authorities. Where vessels are of such size and power that they can keep sailing to fixed schedules with little disruption from bad weather it becomes possible to have firm timetabling, guaranteed berthage and access to specialist equipment. Something similar became possible with the advent of the jumbo jet. However, shipping movements and air traffic always carry a degree of uncertainty. Markets change and ships get diverted. Bad weather closes down airports. Airports and seaports need confirmations, estimated times of arrival and a picture of inbound traffic so that approach speeds can be adjusted to fit in with the movements of other traffic and the availability of docking space. GMSC means that global air and sea traffic can be controlled and traffic harmonized from just a few centres, with any one centre able to back up others in the event of an emergency shut-down. However, the development of such traffic control may be restricted by a lack of international agreement. Sovereign nations earn revenue from control of their air and sea space.

Increasingly nations are following the lead of the United States in requiring detailed documentation of ship's manifests to be forwarded electronically before vessels sail so that cargo can be assessed for entry before it reaches national waters and ports. This level of scrutiny is driven by security needs, but it also makes it possible to project the onward movement of incoming cargoes and the incoming movement of goods to be shipped out.

Skills networks

Those who command the ships at a global level must have Master's tickets that are 'foreign going'. These signify that a person has the skills and experience to navigate ships according to internationally agreed standards and regulations. For coastal and inland waters different countries have their own network of regulations and different certificates for different sizes of boats.

Airline pilots start their flying careers by gaining a private pilot's license (PPL). They may then graduate to flying a variety of light aircraft types and advance to twin engine capability. They may also obtain their instrument rating to allow night flying operations. Once they have gained sufficient flying hours they can be examined and tested for a commercial pilot's licence (CPL). Those bent on a career as a pilot will seek employment with small commercial airlines initially flying as a co-pilot until experienced enough to take command. There follows a progression of learning to fly bigger aircraft, including jet aircraft, in each case obtaining a type rating that only remains valid if sufficient hours are flown in a given period on that type of aircraft. Most commercial airlines ensure that their pilots all gain experience over specific routes and at specific airports as a co-pilot before taking a command role. Advancement comes with seniority and the build-up of hours logged on specific aircraft. Senior pilots act as check pilots to ensure that pilots are familiar with the equipment they are using. Considerable use is made of flight simulators to help pilots learn how to react under particular induced stress situations and emergencies. The skills networks needed in ships and planes are pieced together from practical experience and endlessly revised as new vehicles come into operation, infrastructures are modified and mistakes are made that result in investigations. This is not only at the managerial/officer level, but also at the level of those who form air crew or ship's crew. The careful analysis of these skills provides a pedestal for automation, computerization and robotization.

It becomes a norm for aircraft and ships to switch to automatic pilot. Pilotless missiles fly with precision to over-the-horizon targets.

Unmanned spacecraft journey to planets and their moons and release robotic travellers. Do we need pilots and crew? Computers have better memories and faster calculating skills than humans. Robots can be taught to mimic optimum human performance. And expert systems and robots get better at what they do and learn from their mistakes. Increasingly the role of humans on watch is to be there in case the machines make a mistake, but the likelihood of computers making a mistake decreases whereas that of a human making a mistake remains much as it always has. Unlike people, computers do not get absent minded, drowsy or drunk.

Ships and aeroplanes are acquiring self-regulating autonomy. As it were, they become skilled in manoeuvring and aware of themselves and their environment in ways that exceed human sensory capability. Automatic pilot systems on aeroplanes fine tune for level flight with attention to detail that humans cannot match. In ships, water can be pumped between ballast tanks under the control of expert systems to minimize stress on the hull and improve ship stability. The problems that come from marine organisms being shipped from one environment to another as ballast water is taken on or discharged can be dealt with by sensors that indicate what organisms are in the ballast water, and programme suitable treatment to eradicate problem organisms while the vessel is in transit. Ships and aircraft will in time be able to monitor the status of everything on board from stress on the hull to the health of passengers and crew and the condition of cargo. Such information can be supplied not only to the operating company, but also to the owners of cargo, security agencies, insurance firms and regulatory authorities.

Sensors make possible real-time three-dimensional simulations of aeroplanes and landing strips, ships, docks, locks, rivers, canals, cranes and container gantries. Distributed virtual reality makes such simulations available in different places. HyperRealities can be generated that allow human and artificial intelligences to manipulate a situation in symbiosis. It becomes possible for ships and planes to be loaded and unloaded and navigated without anybody actually being on board. Ships and planes can be controlled from land-based communications centres where an on-duty commodore could keep a weather eye on a fleet of ships or planes through a watch. That is not to say that human intelligence is redundant. AI is unlikely to acquire the social nuances needed at the captain's table or an air steward's ability to deal with the unusual.

It was near the end of a long flight. The 'fasten seat belts' sign had been switched on. Cabin crew were preparing for landing. With exquisite timing, a young child needed a nappy change. Mother and child headed for the toilet. On arrival back at their seats, the husband looked at his wife's hand and noticed that her wedding ring was missing. The Cabin Steward seeing their consternation enquired whether he could help. On discovering the problem he headed for the toilet and emerged a minute later with a big grin to present the mother with her wedding ring beautifully cleaned and smelling of Elizabeth Arden scent. He had guessed rightly that the ring would have slipped off and be wrapped up in the dirty nappy stuffed into the waste receptacle. (*Travellers' Tales*)

Regulatory networks

At the global level, international waters are regulated by the United Nations Law of the Sea, (UNCLOS) with guidance from the International Maritime Organization (IMO). Another United Nations agency, the International Civil Aviation Organization (ICAO) provides a like service for international aviation and oversees the international agreements that govern the conventions of international civil aviation. The International Air Transport Association (IATA) once had the power to control the fares airlines charged and still acts as the international clearing house for airlines to settle their accounts with each other. Passengers may travel on tickets (physical or electronic) issued by one carrier but involving other carriers as well. This allows through-ticketing and interlining with each airline receiving its rightful share of the fare according to set formulae. IATA therefore collects a huge amount of data and can monitor global travel patterns with precision.

At the national level countries have sovereign jurisdiction over their airspace, but there is no agreement over the extent of this space. Nations have shot down high flying aircraft they deemed to be violating their airspace. Most nations, however, can only use diplomatic pressure to dissuade spy flights as they do not have the means to shoot them down. Satellites can overfly any territory and see and eavesdrop on what is happening below. This has given rise to international space law as a specialist branch of international law.

Coastal states have sovereign jurisdiction over their territorial waters and a 200 nautical mile exclusive economic zone (EEZ). Monitoring the activities of foreign vessels in passage through EEZs may be necessary to ensure that there is no poaching of resources or flouting of agreed fishing quotas by vessels operating under licence. This requires regular reporting, accurate entries in log books and visual surveillance by patrolling aircraft besides any physical tallying of catch as it is landed at facilities provided by the controlling state. It is difficult to monitor fishing or seabed harvesting of minerals with exactness without on-the-spot inspections.

At the urban level national legislation is sometimes used to limit the height below which aircraft may not fly for reasons of safety and social well-being. No-fly zones may be imposed around specific places such as presidential palaces and conversely flight corridors may be allowed for low-level flights that include the operations of short take-off and landing (STOL) aircraft using downtown airports and helicopters alighting on helipads on buildings.

Communications networks

The traveller sat back in the taxi as it headed out of Tokyo to the airport. He reflected on his recent experiences staying at a Japanese hotel and the high quality of service he had received. He spoke no Japanese and the signage along the route meant nothing to him, except those that had an icon of an aeroplane and underneath in English 'Narita airport'. It was sufficient for him to realize after a while that the taxi was going in the opposite direction. Angrily he tried to find out what the taxi driver was up to, pointing at his watch and saying 'Narita' and waving his arms and making aeroplane noises. The taxi driver handed him a telephone and indicated he should speak into it. 'Hello' the traveller said tentatively. A voice answered 'Do you speak English?' When he affirmed that he did, the speaker explained. 'This is a translation service. Your taxi driver asks you not to worry. The hotel you stayed at has phoned him to say that you left your passport at the desk. He is returning to pick it up and has checked your flight. It is a little delayed and there will be no problem in getting you to the airport on time.' Just at that moment the hotel came into sight and standing outside was the receptionist holding a tray on which lay the errant passport. (Travellers' Tales)

Global transport brings special communication problems that come from the multiplicity of languages spoken around the world. The principal international language for civil aviation and sea transport is English. Where other languages are used, English tends to be the default language. For those involved in air and water traffic there are conventions to minimize misunderstanding. All letters in the alphabet have a word association that allows words to be spelled out. The registration letters ATCF would be transmitted as 'Alpha' 'Tango' 'Charlie' 'Foxtrot'. Communications between traffic control and pilots usually involve the recipients of a message repeating back the instructions and important messages being duplicated in different media.

Passengers can be bewildered and frustrated finding their way through the sprawl of large international airports where their native language is not spoken. Self-explicit signs can help, but most communication depends on language. What is beginning to make a difference are teletranslation and portable computer-based translation systems. Although in their infancy, they improve (O'Hagan and Ashworth, 2002) and it will become possible for taxi drivers and their passengers to communicate directly with each other even though they are speaking in different languages. The instruments used for such interpretation could be a hand-held translator, a mobile phone or something like a portable stereo headset that transmitted and received by radio. A new generation of airport signage would be able to speak directly to people walking past who held electronic tickets that were also RFID tags. From this they would know what language the individual spoke and what flight they were on. The signage would be able to guide them through the airport, keep them informed about their flight and help them with information. Such signage and surveillance systems would be able to track and organize the movements of people as they came together for a particular flight to make boarding and security a more efficient and benign process.

Such signage systems communicate at the first order of meaning, but when people are confused, frightened, feeling ill or in some way distressed they want some second order communications of comfort and understanding. In transport, as increasingly in management in general, we seem to be at a stage where we seek to make all communications, even human face-to-face communications, function at the first order of meaning in a search for efficiency. But there are always two levels of meaning. Messages that are intended to give clear unambiguous orders can sound authoritarian. There is a danger that we dehumanize the communications in transport, especially at the global level where language problems and security checks introduce extra levels of stress.

Auxiliary services providers (ASPs)

To travel through an international airport is to realize the number and variety of auxiliary services involved in global aviation. There are customs, immigration, ticketing, banking and insurance services. Some of these are essential, like the fuelling services for planes, the lavatories, the cleaning services and the ASPs that wait in the wings in case they are needed, such as the fire and security services. Some, like the concession shops, bars, restaurants and flight clubs seek to add value to transport.

All these networks are linked by a combination of human and IT communications links. Global travel needs global reservation systems. All major airlines and subsidiary carriers together with travel agents use computer-based booking systems. These allow for the input of request details for air travel and associated travel services. Nearly all scheduled airline services in the world are held within such databases, including changes that are foreshadowed in the future. They make possible the matching of demand and supply almost in real time. They now encourage individuals to go onto the internet to make their own reservations and pay electronically. The savings to the airlines of this self-administered reservation system are considerable, but the change is at the expense of travel agents, which now have to charge customers directly for their advice rather than receiving a percentage of the fare from the airlines involved.

Transport services providers (TSPs)

The numbers who take to the air in their own aeroplanes or venture onto the high seas in small boats are relatively small. Most people when they travel by air or send freight by sea use transport services provided by shipping companies, airlines, travel companies and carriers.

A TSP such as an airline is essentially a communications network. It may own a fleet of aircraft and so have a traffic network and an infrastructure of offices, but these could be rented. The existence of the TSP ultimately depends on its ability to sell a transport service.

Until recently that was done through face-to-face links between company representatives and customers at sales outlets and airports or by telephone. This process is now moving on to the internet and becoming automated. When someone goes onto the web and buys an airline ticket that information is distributed to the whole service network immediately. People use the internet to hunt for the transport services they need and transport services increasingly use the internet to hunt for customers. The way people use the internet is stored in the data bases of large companies and such information is sold and cross-referenced. The fact that a person buys books on a particular country makes them a potential customer for travel to that country and is useful information to a bookseller with links to a travel agent.

International civil aviation still stands outside the WTO. Air services are still largely conducted on a bilateral basis and many airlines still have majority national ownership requirements. However, governments have had to allow the formation of airline alliance networks together with their associated loyalty schemes especially where carriers in their own country could not survive otherwise in the competitive global market. The trend is towards interlocking shareholding. Eventually there could be only a small handful of global alliances. This will also encourage standardization of service conditions. Instead of individual airlines competing head to head on service quality, whole alliances will compete with each other.

Seeking space

Will bulk carriers become behemoths gliding silently beneath the oceans, sensing each other by radio, controlled by some on-shore communications centre and devoid of any bridge, portholes or living quarters because there is no one on board? They would be economical. With no one on board to have an accident, drown or be killed by pirates they would be safer.

The trend in global water transport is to bigger more manoeuverable ships with more effective propulsion, mooring and cargo-handling systems. Similarly, the logistics of global air transport point to bigger, more fuel efficient planes, flying further and specializing in handling different kinds of cargo and passengers. Economy class could be further economized by packing people horizontally with a pill before take-off to ensure a good sleep through the flight and freedom from jet lag. For those who wanted to fly for fun there would be restaurants, theatres,

Figure 8.5   No frills airline

These are not normally part of large alliances. Note the web address on the side. Courtesy easyJet Airline Company Limited.

crèches, gymnasia and showers. Aeroplanes could hold thousands of passengers, with no cockpit for terrorists to take charge of because no one on board need be involved in flying the plane.

Sea and air transport have always pushed at the edges of known territory, seeking new routes and new places for travel and trade. Some time in this new century space will become the new fractal level of transport where the nodes are satellites, space stations and planetary space ports and the links are made by rockets. Where jet engines opened up the stratosphere, rocket propulsion opens up the possibilities of travel outside the earth's atmosphere. This could also affect existing levels of global transport. New global networks could come into existence with nodes in the form of rocket ports and routes that depended on the direction of the earth's rotation. Rocket-boosted craft would go straight up until they broke through the earth's atmosphere and then wait for the earth to spin below to bring them close to their destination where they would then land like conventional aircraft.

For sea transport the new frontier lies below the surface of the seas in the unexplored depths of the oceans. With no crew to consider, bulk carriers could be submersibles. Being no longer dependent on human vision for navigation, electronic vision underwater would allow vessels to move with more fuel efficiency, greater stability and on multiple levels like aeroplanes.

9

Driven: land transport by road and rail

When I overtake you on the road, there is nothing personal in it. (Billy Connolly on his 2004 World Tour)

Achieving environmentally sustainable transport doesn't mean less transport than we have today, but it certainly means different transport than it would be under projected trends. Passenger transport would require a substantial shift to public transport, including innovative mobility services with access to car-based mobility and more than a doubling of rail passenger transport over the next two decades. Achieving EST would require that freight transport volumes by rail and inland shipping would have to be nearly doubled from 1995 to 2030, while through technological progress, also growth for road freight of more than 50 per cent is anticipated. (Wiederkehr, 2002)

Introduction

We cannot fly. Few of us can even jump our own height. We can swim, but not far or fast. We are land animals adapted over time to walking warily in grasslands with our heads up and hands free to carry or fight or both. Our journeys begin and end on land. At the urban level, land transport is a daily business as we commute to work or school, go shopping, play sports and walk the dog. It is possible to circumnavigate the world by water and air but not by land. Land transport can be only a part of global transport. The focus in this chapter is on the national and urban levels.

Land transport infrastructures are rail and road networks. In railways, the different NETS are closely identified with each other and it is possible to think of them forming a single system. Railways are for trains and trains for railways and railways have their own regulatory systems and may be owned by a railway company. The road NETS, however, are only loosely linked and serve a wide variety of purposes. Roads are for people and animals as well as cars, trucks, buses and trams.

The fractal levels of land transport networks

Since early times land transport systems have made regional trade possible at the continental level. By 3500 BC, the Persian Gulf was linked to the Mediterranean by a track that was later extended to the Aegean to become the Persian Royal Road. In the second millennium BC, the 'amber' routes linked the Mediterranean to the Baltic and Italy to Spain and by 200 BC, the silk route linked China to India and Europe. Just as the great maritime empires were built on networks of water transport, so the great land empires of China, India, Persia and the Incas were made possible by road networks at the regional level.

Rail networks also developed at the continental level. Russia in its days of empire under the tsars built the trans-Siberian railway to hold its European and Asian halves together. The United States gained its present shape by building the transcontinental railways that joined states on the east coast to those on the west. Cecil Rhodes dreamed of a Cape to Cairo railway that would give Africa to the British.

Modern day supranational terrestrial transport can be found in the transcontinental road and rail systems that link the countries of the European Union and in the Pacific Highway that links the Americas. The completion of the Mauritanian section of the trans-Saharan Highway means that, apart from crossing the straits of Gibraltar, it is now possible to drive on metalled roads from northern Europe to Dakar on the southern edge of the Sahara.

At the national level interurban road and rail networks link urban centres. Road and rail compete for passengers with national air transport systems. On a city centre to city centre basis for distances up to 500 kilometres, direct high speed express trains are probably more efficient than aeroplanes; above that distance air transport tends to become more effective. However, most interurban freight and people traffic is by high speed sealed-surface road networks. In developed countries, these are the expressways, freeways, motorways, *autobahns* and *autostradas* where traffic flows fast and can only stop at designated rest areas. Slow traffic such as pedestrians, cyclists and local bus services is not permitted and these roads do not provide direct access to adjacent property.

Urban road transport includes the roads that link small towns and villages and serve rural areas. This level begins where motorways with their special regulations and restrictions end and the roadside becomes permeable, allowing legal access to property. Where motorways link cities and towns, public roads link buildings. This is the network that can be used by the public in general. It is the level that gives access to the places with an address where people live, play and work and which is the subject of the next chapter.

There is also a level of footpaths, tracks and rights of way. These are intended for people and animals and not for modern motor vehicles and may be private or public. The rationale that brought them into use may have disappeared, but the right to use them is cherished by ramblers. They are remnants of earlier road and rail networks.

The urban level in railways is that of commuter train services and elevated and underground railways. Like public roads, railways also link towns and villages, but such links have tended to die out with the growth of motorized traffic. Many of the routes taken by railways that served rural areas and small towns in the past have now become footpaths, horse-riding and cycle paths. Some urban areas that retained these old rail transport corridors have reintroduced modern light rail tram services linking city centres to distant suburbs or urban renewal sites such as in London's docklands.

Land transport infrastructures

In contrast to water and air transport, where the main infrastructure investment has been in the nodes, in terrestrial transport it has been in the construction of the links.

City stations are the hubs of railway infrastructures, but every minor station or halt, every line intersection and every signalling point are nodes because they are places at which trains can change direction or stop and start. Interurban express train services may have their own tracks with only a few nodes to interrupt the flow of traffic. Where they use a common track with other rail services the minor nodes are suppressed. Express trains are express trains because they travel directly at speed between main stations without stopping.

Similarly, motorways are designed to allow the free flow of fast traffic between the places which give access to and from urban road networks. These ramps are the nodes of motorway networks and spaghetti junctions their hubs. At the urban road level every stop and start interruption, every zebra crossing or intersection, every house along a street, every building that has an address and every site that can be accessed is a potential infrastructural node. Some of these, like traffic lights, apply to all road traffic; others, like bus stops, apply only to a particular kind of transport service.

It is the urban road, not the motorway or the railway that gives public access to the site level of transport. Railways may have spur lines and sidings to some factories and mines and canals may have docks that link to special sites, but it is having an address on a road that legitimizes the presence of a person or an organization in a country. An address directs people to where someone or something is. An address is where, in the last analysis, people live or work, where, from the perspective of transport and communications, they come from and go to. People without an address and so 'of no fixed abode' are called vagrant and regarded with suspicion.

People and animals can cross open country without an infrastructure, but the natural environment provides obstacles and lines of least resistance through them that channel human traffic. This creates natural networks of paths and routes that over time become surfaced roads engineered for wheeled vehicles. People always want better roads for their chariots, wagons and stagecoaches and old roads get adapted to new traffic. Today we have reached an extreme where we find pedestrians pushing prams across public roads used by giant trailer trucks. It is the motorways that are specifically designed for motorized transport. With two or more lanes in each direction separated by a median strip, no crossroads, no steep slopes, no sharp corners and no pedestrians or animals, traffic can flow smoothly at high speeds.

The idea of building a high speed network of roads specifically for motorized vehicles goes back to the *autostradas* of Italy in the 1920s and the *autobahns* of Germany in the 1930s. Adoption by other countries came after the Second World War. The massive scale of this new transport infrastructure meant that many years were spent in planning, seeking resources and authorizations and acquiring land before their actual construction. For most countries it was not until the 1960s that high speed interurban arterial road infrastructures began to carve cross-country links to connect with retrofitted urban motorways.

Much railroad infrastructure is, by comparison to roadways, underutilized. Countries such as France, Germany and Japan have tried to divert terrestrial traffic from roads to rail by building high speed track devoid of sharp curves or steep inclines. Adverts have appeared in different countries with the simple message 'Why get stuck in traffic jams when there is a smooth ride by rail available?' But such inducements have not changed the underlying trend. People want motor cars and the more governments respond by building road infrastructure, the more people buy their own set of wheels.

A possible solution is to change the infrastructure. High speed interurban networks could be re-engineered so that it was the road that moved the vehicle and the infrastructure that controlled traffic. Today, vehicles jockey with each other, often at high speed, to get to their destination. Traffic spreads out to avoid the risk of crashes. A road-driven system would pack cars going in the same direction in platoons locked together electronically and moving safely in unison at speeds which could be two to three times that which freeways currently permit. Exiting the speedways, automobiles would re-enter the public road network to face the congestion of city streets, and the human driver would resume control.

Traffic networks

Freeways are restricted to high speed motorized traffic, but public roads are used by cars, trucks, buses, cyclists, pedestrians, equestrians and, depending on the country, anything from llamas, sheep and cattle to camels and elephants. These different kinds of traffic move at different speeds, with different intermittency, according to different regulations and involve different kinds of driving skills. This is a recipe for traffic chaos. A logical solution is grade separation. This means that each kind of traffic has its own network with its own infrastructure. To some extent, this exists in the form of pedestrian pathways, cycling lanes, bus lanes and drovers' lanes, but the great traffic problems of the world are in the large conurbations where competition for space is critical. Traffic restraints are being introduced. Singapore has auto-

Figure 9.1   Kota Kinabalu elevated walkways

Pedestrian paths are separated from vehicular traffic by the construction of elevated walkways over streets and connecting buildings at first or second floor levels.

mated road congestion pricing and London promotes public transport and has created central city no-go areas for cars. Despite such measures, private motoring remains the preferred mode of travel at the urban and interurban levels. Motorized traffic uses 80 per cent of the energy involved in transport and continues to grow dramatically. On current trends there will be over a billion automobiles by 2025 (WRI, 1998). Automobiles are by far the most inefficient, expensive and dangerous form of transport and a major factor in global warming. Yet, few people reading this will not have a car and, while they may in principle approve of reducing automobile traffic, they would not want to become dependent on public transport.

In an episteme of growing democracy, wealth and mobility, governments come under pressure to satisfy the public lust for cars by building infrastructure and keeping the cost of motoring down.

Virtually every country ignores the full costs of motoring that include policing, emergency services, deaths and disability and damage to the environment. In so doing they subsidize motorized transport. Democratic governments would have serious difficulty facing an electorate if they were to apply a real user pays policy to motorists that would radically readdress the costs to a nation of unrestricted motoring. If they did, not only would they suffer the opprobrium of motorists, but transport costs would soar and the economy would suffer as it competed with other countries that continued to subsidize oil-based transport.

A first step towards reducing pollution from motorized vehicles would be to change the means of propulsion from total reliance upon fossil carbon fuels. Hybrid powered vehicles are gaining in popularity. At slow speeds in congested inner cities involving many stop–go situations, electric motors could cut in. Some energy could be recaptured during braking. On the open road in free-flow conditions, motive power could revert to diesel or petrol. Other systems are possible such as compressed air driven cars and cars that use hydrogen powered fuel cells. Water is the by-product of burning hydrogen. There is no release of greenhouse gases, but manufacture of the hydrogen fuel would need to be effected by renewable energy sources.

Skills networks

The people who drive trains and pilot planes and ships are a relatively small group of dedicated professionals who undergo extensive training and testing over years of progressive advancement through supervised practice. With the exception of Saudi Arabia, where women are barred from driving, any adult can drive a car, provided they can pass a driving test. Although driving a motor vehicle is a complex and demanding skill, few people fail to get a licence although it may take them several attempts to do so. Unless they are elderly, they may not face health checks except perhaps for their eyesight. Young drivers are not tested for emotional maturity. Old drivers are not tested for their speed of reaction. It is hardly surprising that going by road is far more dangerous than going by ship, plane or rail. There are many people with low skill levels driving automobiles and they kill far more people than terrorists.

Regulatory networks

Railways are easy to regulate because they are dedicated to one kind of traffic, but roads have multiple kinds of traffic which involve a wide spectrum of users. Children and elderly and disabled people use the same roads as professional bus and truck drivers and cyclists. Every country has a highway code of some kind that specifies how road users should behave, but this is only seriously tested when people seek a licence to drive a motorized vehicle. In some countries pedestrians can be prosecuted for jay walking and there are mandatory requirements for cyclists to wear safety helmets, but in general regulations governing the fitness of the people who use roads are not enforced as rigorously as they are in rail transport, or for that matter in air and water transport. Where regulations are carefully enforced in road transport is where they relate to vehicles. Many countries test motorized vehicles on a regular basis to ensure they comply with a raft of regulations as to the condition of such things as horns, tyres, brakes, steering, couplings, lights and rear vision mirrors, but there is little attention to non-motorized vehicles such as bicycles, scooters, prams and skateboards or the people who choose to use them.

A large part of road transport costs are in policing and testing the regulatory network and in the emergency services, hospitalization, rehabilitation, loss of work and disabilities that are consequent on people's failure to observe regulations. These costs could be reduced with the application of IT. People begin to carry GPS devices. These could have built into them the regulations that apply to the section of road they are using. If cars drove themselves, they could monitor their own safety status and refuse to move if they needed repairs or renewal. Clever clothes made with nanotechnology could similarly look after the welfare and safety on the road of the people wearing them.

Auxillary service providers (ASPs)

As with transport by water and air, terrestrial transport is made possible with ASPs that provide emergency services, policing, fuel, food, shelter and so on. Road transport in particular gives access to a wide variety of services that range from shops, hotels and restaurants to parks, libraries and schools. Perhaps the most important development taking place in terrestrial ASPs comes about with mobile telephony. This communications network is becoming the principal means for accessing services of every kind.

Communications networks

Mobile phones are becoming mobile communication devices not only for text and image but to access radio, television and the internet as well as the telephone services. In the future this will become the kind of wearable IT described in Chapter 7 that allows people to go to places as full-bodied tactile telepresences. The link between telephonic addresses and road node addresses is weakening. In the future telephonic addresses could be based on people rather than places. With an RFID tag that linked their telephone number to their internet address, their passport number and their social security number, they would be floating nodes in a transport environment rather than fixed nodes at the ends and beginnings of transport.

> The ticket was booked on a website, a hole-in-the-wall machine issued it and the unmanned barrier opened when it was fed through. In the early morning the traveller found his seat on the train without problem and unfolded his newspaper. Every few minutes a recorded voice announced that anyone who was seated in a train or compartment where they were not authorized would be rigorously prosecuted. It made the traveller uneasy and he kept checking that he was in the right seat on the right train, but everything seemed in order. The train left on time and arrived on time and the traveller exited through another automatic turnstile after a smooth journey.
>
> Later in the day the traveller arrived back at the station for the return journey. A milling crowd of angry people were gathered around the departure board which announced a bewildering variety of trains to his destination, all of which read 'delayed'. What on earth was going on he wondered, but there was no one to explain. Finally, he found a friendly old railwayman having a cup of tea in a quiet cubby hole who told him that a train had broken down on the main line and all the following traffic had stacked up behind it, hence all the 'delayed' notices issued by the computerized system. The old man pointed out the platform the traveller needed and suggested he got on the first train that

came that was going to his destination. 'But what if it is not the train I am booked on?' asked the traveller remembering the dire warnings. The old man laughed and wished him luck.

A train arrived that seemed to be going to his destination and the traveller boarded. Then he saw why he needed luck. The train had been delayed for many hours and was crammed with football fans. Many were drunk. They were singing, dancing and arguing aggressively. Some were conducting an impromptu game of football down the corridors. Others lay on the floor unconscious in their vomit. Women cowered in their seats protectively hugging their children. The toilets were hideous. The speaker system carried no threats of prosecution for sitting in the wrong seat. Instead a voice kept announcing to cheers from the football fans, that while there was no more food on the train, there was still an abundance of alcoholic beverages available.

The traveller completed the nightmare journey and exited through the automatic barriers. He realized that the only member of the railway staff he had met was the old man and glanced back at the engine that had pulled them in, but could see no sign of a driver. Had anybody been in charge? (*Travellers' Tales*)

Is this the future of rail travel? People driven by automated systems from which the last humans have been made redundant? As long as people do as they are told and as long as everything works as it should, as was the case in the first half of the traveller's journey, such a system is perfectly feasible in the relatively closed systems of rail, air and sea transport. But sometimes things go wrong and people have emotions and needs and get excited, angry, exuberant, drunk, disorderly and sick.

Mass transit systems are steadily replacing the staff who interact with passengers, such as ticket collectors, with automated systems. In so doing they are losing the human communications that are necessary in transport, especially when things go wrong, as ultimately they always do. That some interaction, such as checking a ticket, has been automated does not eliminate the inter-human communication functions that ticket collectors carried out when they responded to the multitude of questions that people asked while they were getting their tickets punched. Since these are varied, and often personal and not clearly articulated, they are not readily computable.

It may well be that this is one of the reasons for the popularity of the private motor car. It gives people door-to-door transport and puts them in charge of their own transport destiny. It is the driver of a car who decides where to go, when to go, what route to take and what speed to go at. Cars allow people to express their individuality. They are vehicles of communication as well as transport.

Driving as communicating

Trains, aeroplanes and ships are means of communicating in the sense that we take them to get to places in order to talk to people or go to the cinema or to deliver letters and newspapers. What we are considering here is a car as a medium of communication that can be used to express its owner's personality, feelings and social relationships. People buy cars as they buy clothes or cosmetics because they think it improves the way they look and says to the world at large that they are rich or sporty or sexy. They can also express in the way they drive their anger, urgency, rudeness, courtesy and consideration.

Communications by car take place within a car, between cars (and other road vehicles) and between a car and the road infrastructure. As with all communications, there are two orders of signification.

Intra-car communication at the first order is done by the driver interacting directly with the interface controls and instrument panels of the car to get it to do what the driver wants. The second order is that of the driver talking or shouting, overtly or silently at the car or being pleased or irritated with the way the car feels and looks.

The communications between vehicles that are within sight of each other and able to influence each other's movements and respond to each other's signals is linked to communications between vehicles and the road infrastructure that is in line of sight. The two kinds of communication happen in the same time and space and have to be coordinated. First order communications is vehicles reacting to the signals and movements of other vehicles and to signals and information from highway signage according to a highway code. Second order communications come in the speed, timing and spacing of signals and manoeuvres and in supplemental communications from the heavy use of horns or interpersonal communications between the drivers by a wave of the hand in acknowledgement of a courtesy or as they lean out of their vehicles to gesticulate, shout imprecations and give finger signals.

Intra-traffic communications are between all the road users that constitute nodes in a traffic network including those that are out of sight of each other. The term refers to all the traffic across all the linked roads in an infrastructure and could therefore be at a national level. First order communications at this level, as at the inter-car level, is in the way traffic communicates according to the highway code in the expectation that if they do, the system as a whole will work. To a surprising degree, despite second order quirkiness, and with the help of traffic advisory broadcasts, it does, but not as well as in railway, shipping and aviation intra-traffic communications, where traffic control centres have an overall picture of traffic on a particular route.

Railway traffic is organized so that trains depart and arrive at fixed times, at fixed points on fixed platforms according to fixed schedules. While there can be failures in meeting them, schedules exist. In most road traffic, apart from bus services, they do not. Vehicles enter and leave traffic networks as and when their drivers want and the drivers bring with them moods and attitudes that are a source of second order communications. Unlike intra- and inter-car communications, second order communications at the intra-traffic level are not synchronous. What happens in one incident can set a mood that affects other traffic encounters and can ripple through a traffic network. Traffic police say they can sense collective moods in traffic.

Road traffic endlessly seeks organization and organization is always emergent, but never fully established because the elements in road traffic are always changing. There are theories in communications studies that seek to explain how organization emerges through interaction in a common field of endeavour (Everett, 1994; Taylor *et al*, 1996). Communicative behaviour of individual vehicles in proximity with each other in time impacts on the traffic network as a whole. In turn the cumulative communication behaviour of traffic as a whole over time creates a traffic culture that then begins to impact on each individual driver making the driver conform to a norm of behaviour that has been created by the traffic itself. This may not be the same as the official norm of expected behaviour set down in a highway code.

Weick (1979) in his theory of organizing postulates a pattern of interaction that begins with an *act* of communicative behaviour by an individual. When an act leads to a response from another individual this is an interact. When an interact leads to a further reaction by the first person it forms a double interact which Weick believes is the basis of all oganizing behaviour. Imagine driving along above the speed limit and coming up to a car that is driving below the speed limit. You sit on its tail looking to overtake. This is an act. The other car notices and speeds up a bit so that it is slightly above the speed limit. This is an interact. You decide you would have to go too fast to overtake so you drop your speed a little, drop back a bit and drive behind the other car. This completes a double interact that has one car driving faster and the other slower than before, but results in both cars driving slightly above the speed limit. This kind of double interact is continually happening right across the public road network as traffic organizes its relationships. The results of each double interact will of course vary, but if on average the outcome of each double interact is that the cars concerned go a little faster, then the self-organizing nature of road traffic will mean that traffic as a whole will begin to flow faster. This may happen sporadically as when a number of high spirited people flood on to the roads on a public holiday or it may become so normal as to constitute a *de facto* traffic behaviour that is different from the *de jure* behaviour that is expected on the road. It is not uncommon to be in a stream of traffic that as a whole is moving considerably above the posted speed limit for a particular sector of road. Countries may seek to force traffic to adhere to the legal speed limits by advertising and policing campaigns, but sometimes authorities accept the change and alter the posted speed limits to match the reality of a traffic culture.

Part of Shakespeare's style of writing was to invent new words and new ways of putting things. Imagine if when he was doing so his quill pen suddenly said to him 'Hey, there is no such word as that.' We do not write with quill pens any more, we use computers and they have spelling checkers that do let us know if the word we use is not in the dictionary or is incorrectly used. Cars are no longer submissive servants that do exactly what they are told. They have become back-seat drivers that tell us to belt up, that we have left the lights on or that we are going too fast. They seek to persuade drivers to revert to a norm of behaviour based on first order observance of the highway code.

The computers in cars can be linked to telecommunications. Drivers no longer need wayfinding skills. They type in their destination and the car tells them how to get there. If on the way the car learns of any traffic problems it can re-route. Nor do drivers need car management skills. A vehicle can continuously monitor the condition of its vital systems, and warn the driver of dwindling fuel supplies, fading brakes, deflating tyres, cracked wheels, engine overheating, loss of oil pressure and faulty lights. Linking this to telecommunications means a car could check itself in for a service (with the garage's computer). If an intoxicated person is in the driver's seat, the car could be programmed not to start. If, from monitoring the eye movements of its driver, a car detects the onset of fatigue, it could advise the driver to pull over and rest. If the driver ignores such warnings, the vehicle may automatically start flashing hazard lights to warn other motorists and enforcement agencies. Cars can pass without stopping at toll gates by using electronic exchange of information for electronic billing. Road user charges and taxes can be collected by telecommunications and can be calculated on the basis of time of day, position of vehicle, weight of vehicle, and whether the vehicle is correctly licensed to use a particular route or be in a particular zone at a particular time.

We come back to the fact that the growing application of IT in automobiles raises the possibility that cars could drive themselves; that one day they will, like trusty chauffeurs, adjust seats and climate control and ask where we want to go and then take us there without any human intervention. Operating only at the first level of communications, traffic would surely be safer at every level. At intra-traffic levels all automobiles will be in simultaneous communication so that, knowing exactly what the rest of the traffic is doing, a vehicle can speed up and slow down to avoid peaks and surges and allow for dangerous conditions and accidents. It would be an end to driving as an antisocial way of life. But the trouble is, as Jack Kerouac understood, we like driving. It is a way of communicating available to many who are not articulate in other media (Kerouac, 1958).

The future for land transport

As with ships, planes and trains, road traffic could one day drive itself with AI that is either in the vehicle or in the road. But ships, planes and trains and road traffic in freeways do not operate in an environment as complex or as charged with humanity as that of urban road networks. We are inculcated with the paradigm of urban road transport from earliest childhood. We know how to use traffic lights, zebra crossings, and bus and taxi services and how to shop and go to the cinemas and parks that lie alongside roads. We know the paradigm of road use in the round from the multiple perspectives of pedestrians, passengers and drivers and we know it in a deep way from experiences that involve having to manage when things go wrong. We know that on the streets we will meet with unexpected behaviour and inconsistent communications. This is not the kind of environment that computers easily adapt to. The next chapter takes us further into this as it explores on-site transport networks.

10

Walking the walk to talk the talk: pedestrian transport

... sensory-motor controllers (used for walking) may be directly coded in our neural circuitry and be available to other cognitive processes such as language interpretation, and more relevantly may ground the semantic and grammatical structure of the well known linguistic notion (that walking the walk is like talking the talk). (Narayanan, 1997)

As I walked along the Bois de Boulogne with an independent air, you could hear the girls declare, he must be a millionaire. (Music hall song 'The man who broke the bank at Monte Carlo')

Introduction

A container truck backs up against the storage depot of a supermarket. It has carried tons of different products at high speed along a motorway network and then at a slower speed through an urban road network to this site. This is state-of-the-art technological man-machine terrestrial transport efficiency. Only one human was involved in moving a large quantity of freight over a long distance quickly. Now a number of humans will be involved with far less efficiency as the truck is unloaded and goods are moved short distances into storage, unpacked, bar-coded and shelved. In developed countries, this on-site transport will be done with the aid of conveyer belts, pallets and fork lifts, but around the world much of this work will still be done, as it has been for thousands of years, by humans lifting things and humping them on their backs or carrying them in their arms.

Next, shoppers do similar transporting activities in reverse as they lift things off shelves and cart them to a checkout counter, bag them, trolley them out to their cars and pack them in the boot. Once they are in their cars, shoppers shift back into the high technology transport mode they share with container trucks as they move themselves, their goods and their children from the supermarket to their homes via an urban road network. Then, once again, the shopping is carried into the house and stored, things are put into the fridge and taken out, and people walk and carry, as they cook and carry the baby and play and work. They revert at this, the level of the building, to ways of transport that are as old as the body that evolved in response to them.

This chapter is about pedestrian transport; people taking themselves somewhere on their own two feet and carrying things whether it is on their backs or in a bag or pocket and the way they communicate as they do this. It takes us from the fractal level of urban roads and streets, which pedestrians share with cars and trucks and trams, to the fractal levels of sites, spaces and surfaces, where people are still the main means of transport. How has IT impacted on these levels of transportation? Compared to other levels of transport the answer has to be, so far, very little. In fact, in developed countries, back problems and increasing obesity mean that pedestrian-based transport is less efficient than it used to be. If we are concerned with efficiency this is where developments in transport are needed in the future.

Walking talking transport communications

Supermarkets use barcodes and computers to automate the flow of quantitative first order information. When a customer goes through checkout the store knows exactly what the customer has taken from the shelves and can then work out how well an item is selling and when to order more of a product. Alongside such IT systems, people communicate with each other as they have done over the centuries. They navigate trolleys past each other by such traffic protocols as 'Excuse

Figure 10.1   An infrastructure with two uses

The infrastructure of the railway set up to carry large loads quickly over long distances unintentionally serves as an infrastructure for pedestrians carrying small loads, slowly over short distances.

me', 'Can I get past, please?' and 'Thank you' and as people shopping have always done, they greet friends, gossip, strike up conversations and ask questions. Speaking is to communications what walking is to transport and the two are closely allied. Speech is transmitted by sound waves and the acoustic range of the human voice is limited. Speech, even if it is technically amplified, is not readily decipherable much above 100 metres. When people want to talk clearly and easily they move towards each other. Then they either stand so that their bodies are oriented towards each other or walk with each other while they talk. As the quotation at the front of this chapter from a researcher at Berkeley suggests, there is a possibility that the same neurological processes are involved in walking and talking.

Just how close do people like to get to talk? It depends on what is being said and to whom. To say something to a large group of people means standing well in front of them and speaking loudly. To whisper an endearment to a significant other means getting close enough to touch them, look deep in their eyes and smell them. Hall (1959) introduced the term 'proxemics' to describe the way people use space to communicate. He introduced the term 'informal space' for the personal territory that people seem to carry around with them that measures how close they allow other people to get. When we walk and talk with someone we are endlessly adjusting the distance between us to what we feel comfortable with and this can vary from person to person, occasion to occasion, culture to culture and, as the content of communication changes, moment to moment.

Hall also noted how we use 'fixed space' by which he meant the shape and size of rooms and 'semi-fixed-feature space' by which he meant using adjustable things such as furnishings to communicate. For example, you might choose to meet someone at a restaurant for an intimate conversation and reserve a table where you cannot be overheard. First you and your guest use the urban network of roads to get to the restaurant. You open the door and are met by a member of staff who takes you to your table using a pedestrian transport infrastructure that radiates from the entrance area to each table. Another set of routes for the waiters radiates from the kitchen to each table. However, you do not just come to eat. A vast amount of transport has as its objective bringing people together in homes, offices, schools and conventions in order to be able to talk with each other.

We do not just talk. Imagine that you have invited people to dine with you who are extroverts and who make an entrance that has everybody in the restaurant watching. Your guests are communicating who they want people to think they are with the way they are dressed, the accessories they are wearing, their haircut, their gestures, their body language, their scent and *the way they walk*.

Birdwhistell developed the theory of kinesics (1970) which sought to explain the meaning in body movements. Like other researchers in non-verbal communications (Ekman and Friesen, 1975), Birdwhistell was particularly concerned with the relationship between body movements and language that comes with gestures and facial expressions.

A basic axiom in communications is 'One cannot not communicate' (Watzlawick, Beavin and Jackson, 1967). It implies that any perceivable form of behaviour can communicate. So we might say it is axiomatic that 'one cannot walk (or run or swim or take a taxi) and not communicate'. No one has established exactly how we communicate by the way we walk, but to watch people walking and talking together is to realize how closely intertwined and complex transport and communications are in humans. In the old music hall song quoted at the beginning of

Figure 10.2   Charlie Chaplin's tramp

The tramp walked in a way that moved people of all countries and helped to make him the most famous man in the world in the days of silent film. His comical strut was abetted by the way he swung his stick and his baggy trousers flapped. The image he created persists in signage.

the chapter, the singer is saying that the way he swaggers, the way he dresses and the fact that he is on one of the most fashionable streets in the world combine to communicate that he could be a millionaire.

No telecommunications capability can match the bandwidth of human vision and hearing, the sensory variety that includes touch and smell in face-to-face communications and the intricate relationship between transport and communications that is possible at the site level. Think of what happens when a waiter takes an order, uncorks a bottle of wine and serves a dish, or what might be involved in carrying a new piano into a recently decorated drawing room. The most difficult function to automate in a supermarket is that of bagging. The small common-sense decisions such as whether an object will fit into a bag or recognizing that cheeses from the deli do not go with the onions are the most difficult to program.

The fractal level of the site

The fractal level of transport infrastructure below that of the urban network of roads is that of the site. By site is meant the places where people live, work and play that have addresses that people can go to or send things to. In transport they are places of departure and arrival and in some way defined either as buildings or by walls and fences. Access to them is by a door or gate. A site could be a university, a factory, an airport, a dockyard, a house or a shop. Sites are nodes in urban road networks. Open the door to one and it becomes a network in itself in which rooms or dedicated activity areas are the nodes linked horizontally by passages and paths and vertically by stairways and elevators.

Buildings, walls, rooms, corridors, stairs and fences constitute the transport infrastructure at this level. The traffic is normally, but as in the case of factories and farms not always, people getting themselves and the things they are carrying from the urban fractal level of the street to rooms and activity spaces and moving between rooms and activity spaces.

Large sites such as dockyards and airports may have travelators, elevators, escalators, gantries or some kind of tractor or truck to facilitate the on-site movement of goods and people. Sites that handle containers have container-handlers that lift and carry at the macro level, but, in general, transport at the site level is still largely done by people. Often the hardest part of a journey is getting luggage down the stairs and into the car and then out of the car and through the airport, bus station or railway station and then onto a vehicle. Large sites such as airports and seaports changed enormously in the 1960s, but the interiors of the houses people live in have changed little in terms of the transport functions within them.

The fractal level of the activity space

The fractal level of transport below that of the site is that of the activity space which is typified by rooms. Walk along a passage and enter a room and it switches from being a node in a site network to a network in itself. These are the spaces in a site sectioned off for some dedicated activity such as cooking, sleeping, entertaining, education, office work and selling things. They may be within a building and completely enclosed as a room or partially enclosed as a parking space or swimming pool. The transport infrastructure within such spaces depends on the activity they are used for. The nodes are installations or items of furniture that enable people to go about their business in the room and the links between them are the spaces that people move through to get to them. So a bedroom will have a wardrobe where people can hang their clothes and a bed for them to sleep in and there will be space to move from one to another. A common denominator of all rooms and activity spaces is the doors or gates that allow entry and exit between the room network and the site network

Besides providing a sheltered environment for the activities that go on inside them, these spaces are designed to enable the face-to-face communications that go with a dedicated activity. The communications that go on in a library are different from those in a classroom or a tennis court. Rooms have walls that keep outside noises out and make it possible for everyone in them to hear and be heard.

The fractal level of activity surfaces

Yet a further fractal level of transport can be discerned below that of the activity space in the furniture, work benches, desks, chairs and tables that enable people to do whatever it is that is the purpose of the space they are in. A generic term for this is an activity surface. It may seem pedantic to include the movement that takes place on a table top in a book about transport, but the more we seek to link source to user in seamless transport systems the more we need to think in terms of every fractal level of the transport process. Work benches in factories are where many of the products moved by transport have their origins and where time and motion studies seek to analyse movement at mini level and micro level in order to improve efficiency and allow automation. One day nanofactories will operate at the desktop level and although they might only need microscopic amounts of some rare mineral, the transport issue from mine to factory will still be there.

Figure 10.3   A sewing room

This is an activity space used by a quilter in which we can see an infrastructure with four surfaces and space to move between them. The surfaces as nodes are a swivel chair with castors, a working table, a computer table and an ironing board. The quilter uses the chair as a semi-floating node to move between the activity surfaces in the three fixed nodes. As an activity surface and a transport network in its own right, the working table has semi-fixed nodes in the form of a box of threads, a tray of fabrics and a sewing machine and the quilter moves the threads and fabrics so that they come together with the sewing machine. She also uses scissors, pins and a ruler which have their places on the table along with spare needles and bobbins. A cup of coffee is a node that provides supplemental energy. The quilter uses another surface to communicate. Her laptop shows the design she is working to and gives access to the internet to find and buy fabrics and patterns, take classes and sell her quilts. The telephone supplements the computer and keeps the quilter in touch with fellow quilters working at home in a similar way.

Ports, portals and protocols

The French for door is *porte* which is linked to the Latin *portare* (to carry) which takes us back to the basic root of the term 'transport' (to carry across). The derivations are important because they refer to basic transport and communications functions in the same way. A porter can be found near doors or gates to help people carry things through. A port on the side of a ship is a kind of door that can be opened for carrying things on board or ashore. An im*port*ant person is one who should be allowed through a door because what they have to say is of im*port*, a re*porter* is a person who carries back news or re*ports*. The term has been carried across into IT. Ports on computers are where peripherals can be attached so that information can be passed. A portal is simply a large or grand port such as the imposing entrance to a palace. Airports in their advertising often call themselves the portal to a country or a region. The search engines that give access to the world wide web are called portals.

What seems to be common in the idea of a port or portal is that, regardless of whether it is used of a transport or communications network, it is where a node in one network gives access to another network. The gates where passengers board their aircraft at an airport are the ports where people shift between the network of the airport and a flight network.

The purpose of a network is to allow goods, people and information to flow through them. The purpose of a port or portal is to allow them to move from one network to another, or to stop them. What decides movement through a port is possession of the appropriate protocol. A protocol is the code of conduct to be followed in a network. The term comes from the ancient Greek word for the first page of a manuscript which told what it was about, who it was written by and in what language. It was the basis on which a person decided to buy the book and enter the network of information within it. By the same token when someone buys a ticket to make a flight they accept all the information they never read on the back of the ticket that explains what they will get for their money and how they are expected to behave. When a person passes through the immigration port in an airport that is foreign to them, they are expected to behave in accordance with the laws and customs of that country. The term 'protocol' has been adopted in IT to refer to the rules that govern a particular information network or software system. These days IT does it automatically. We listen as we click on our e-mail icon for a sound like a robot spitting. We call it a 'handshake' as our computer seeks access to the internet. If successful we say we are 'in' the network of the internet.

To get in to the network of a site from an urban network of roads, people look for the door or gateway that constitutes its port of entry. Once they have found it they need the protocol to get through it. Traditionally this amounts to knocking and waiting for someone to come and answer the door. That person will enquire who wants entrance and why and then decide whether or not to admit the individual. Alternatively a person might have a key or swipe card which allows them to do an automatic handshake with the door.

Once people have access to the site network they go to the room or activity space that they are looking for. There will be another door and another protocol that gives entrance to it. Even at the level of furniture, joining a group activity at a table is not done without some form of protocol: 'Do you mind if I join you? Is this place taken?'

Pods and packaging

A pod is some form of container or housing or unit load device that is shaped and adapted to hold what is within. We use the term generically for all the ways in which we package or contain people and things to store and transport them. Thus the term 'pod' could include houses and rooms as well as railway coaches, bus bodies, the hulls of oil tankers, containers, car bodies and the incredible variety of cartons and sacks we find on the shelves of supermarkets down to egg boxes and tea bags. Pods may be designed to nest in each other. This means that besides adapting to what is in them, they are also designed to fit into the pods that in turn hold them. This sounds like the Russian dolls that fit inside each other and to some extent we see this in the dominance of the box shape in packaging. From ship's holds to containers down to individual cubes of powdered soup, things are packed in box-shaped pods that are increasingly standardized to maximize the use of space.

Seats are pods for people. Clusters of seats can be podded in coaches. These can be carried on poles by people and called litters, tied on to the backs of elephants and called howdahs. If they are put on wheels and pulled by horses, or drawn by steam engines along railways or by combustion engines along roads they are called coaches. What they all have in common is that they are boxed spaces with windows, doors and seats.

Figure 10.4 Container shipping

The vessel's shape reflects that of the containers it carries.

We normally think of transport beginning and ending at the level of the site. Those sites are the origins and destinations of goods consigned to be transported and the places from which and to which people travel. Yet the site, room and furniture levels of transportation can themselves be transported. The nesting of seats in rooms in houses can be adapted for transporting people. An aircraft carrier is a floating airport. A liner is a floating hotel. A train or an aeroplane is a moving string of rooms. A car is a small room that moves. A bicycle is a moving seat.

Technological extensions of pedestrian transport

In the last 100 years we have improved the speed with which we can travel from one end of the world to another from a month to a day, but the only improvement to the baggage we carry on such journeys is to add some wheels that are not even powered. In fact, things get worse. The porters, who in the past helped travellers carry their baggage and stow it, have largely disappeared; the distances that have to be traversed

Figure 10.5 A boat in a Bangkok *klong* or canal

The boat has the same interior transport system as a bus, a train or an aeroplane. There is a passage running its length with small passages to seats running off it. It is also similar to the *klongs* it serves which have a main waterway with small inlets giving access to individual houses.

with luggage have increased; seats get smaller as the average size of posteriors gets bigger and now the fear of terrorism adds innumerable check points with primitive packing and unpacking systems.

The US Department of Energy's project on smart communities notes that in the United States alone there are more than 76 million residential buildings and nearly 5 million commercial buildings and that by the year 2010, another 38 million buildings are expected to be constructed. These buildings use one-third of all the energy consumed in the United States. A major, albeit unspecified, part of this is to enable transport at the site, room and furniture levels. (SmartCommunitiesNetwork, 2005)

Buildings are a major source of air pollution and climate change, but solutions could be found in the application of AI. Towards the end of the last century the idea of smart computers in smart rooms in smart buildings in smart blocks began to attract attention. Singapore as an

Figure 10.6   A liner entering Vancouver

The liner looks like one of the skyscrapers in the background lying on its side in the water.

island city state even extended the idea to include smart cities in smart countries.

Terms such as 'smart', 'intelligent', 'automated' and 'electronically enhanced' describe buildings that use IT to improve energy efficiency, security, safety and to provide a rich information environment. These concerns existed in medieval castles where servants would scurry along narrow passageways that ran through the walls themselves, observing through peep holes what the people they served were doing so that they could have meals, fires, baths and clothes ready in intelligent anticipation of the needs of their masters and mistresses. Within intelligent buildings, IT is the servant. Doors recognize people and whether they have the protocols that allow admittance. Lights go on and off as people enter and vacate rooms. Air conditioning adjusts to the tastes of an individual or to suit a group. Walls whisper or show information that the passing person needs and transmit any message they may want to send. Fridges recognize when they are running short of food and order fresh supplies.

McLuhan startled the world in the 1960s with his idea that technologies were an extension of human capabilities (1964). He argued that electrical systems were extensions of the human nervous system, electric lights extended our ability to see, the telephone extended our ability to speak and clothes were an extension of our skin. Taking this metaphor to the field of transport, we can think of wheels as extensions of our legs and bikes, scooters, trolleys and wheelbarrows as technologically extending human transport at the pedestrian level. By the same token we might think of levers, pulleys and shovels as extensions of our arms for lifting. Accordingly, clever clothes would extend musculature to improve people's ability to walk, run, jump, swim and even fly. The clever clothes described in Chapter 6 could lift and carry and move things at an altogether different order of weight and difficulty than any humans have hitherto managed. Tools could become prosthetic devices that were extensions of a person's clever clothing. A crane could be an extension of the hands of the person controlling it.

McLuhan was also the man who saw telecommunications creating a 'global village'. The biggest change at the level of the site and room could come if nanotechnology reduces the need to transport people and freight while at the same time developments in telecommunication made it possible to travel as telepresences far beyond conventional limits.

11

Communications for transport logistics and global supply chains

Communications is key to the efficient functioning of any system, whether it be the distribution system of an organization or the wider supply chain. Excellent communications within a system can be a key source of competitive advantage. (Lambert, Stock and Ellram, 1998)

An army marches on its stomach. (Napoleon Bonaparte)

Introduction

Chapter 3 described how transport, whether by land, water or air, was made possible by the interaction of six different kinds of networks (NETS): infrastructure, traffic, regulations, communications, auxiliary services and skills. The last three chapters have looked at the way each of these networks operate at different fractal levels, which range from the international to the regional, the national to the urban, and the site to activity spaces and even activity surfaces.

We turn now to how people who want transport negotiate the different enabling networks in the different transport modes, at the different fractal levels in order to get people or goods from one place to another as seamlessly as possible. They can of course do this as an individual initiative, as when someone simply gets into their car and goes somewhere. What concerns us in this chapter is when they use a transport services provider who takes over the responsibility. We are not so much concerned with the transport services providers that operate at a single fractal level, such as a taxi service or a ferry, as with global transport services providers that have to integrate operations at different fractal levels in different transport modalities. To manage this, they view the process as a supply chain.

A supply chain consists of the sequence of transport services which link the original producer of a good to its final consumer. Figure 11.1 shows a supply chain network in terms of the fractal levels of transport from site level to global level. The links are arrowed to indicate the direction of flow from upstream procurement to downstream distribution. Flow of product is tracked as it moves from original source

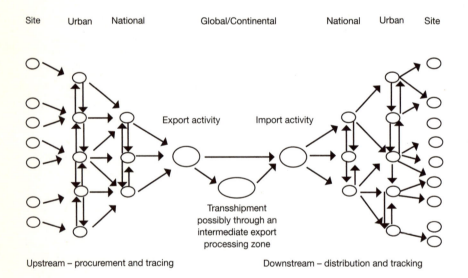

Figure 11.1 The fractal levels of a supply chain from site level to global level

Every link and its nodes requires the six NETS: infrastructure, traffic, regulations, skills, communications and auxiliary services. Any two or more linked nodes in this diagram constitute a supply chain. A shift in fractal level takes place at every node.

to final consumer. Likewise, historical flow is traced back to source. Communication, therefore, flows both ways in the chains of this network. Communications are also critical in what happens at each activity node (by critical is meant that without communications the system cannot function).

Global supply chain management integrates, in an efficient way, all the people involved in the supply chain, including producers, warehouse staff, customs officers, wholesalers, retailers and customers. The aim is to make the supply chain function so that the right merchandise of the right quality is produced and distributed in the right quantities, to the right locations at the right time in a way that minimizes system-wide costs yet meets service level requirements (Mandelbrot, 2006). Supply chain management is not just about minimizing transportation costs or reducing inventories, it seeks system-wide solutions. This includes bypassing intermediate steps where possible. Therefore, not all players necessarily benefit from system improvements. Some supply chains at some stages rely heavily upon labour that may or may not be highly skilled. Elsewhere in the same chain, processing and handling will be intensely automated, requiring deployment of smart and sophisticated technologies. There is, however, a tendency through time for unskilled labour to be replaced by machines and by multi-skilled workers conversant with IT.

NETS can be used to analyse the functions commonly associated with logistics: transportation, warehousing, sourcing and procuring, making or transforming, distribution and customer relations. Delve into any one of these and NETS are in operation. To take warehousing as an example: there is an infrastructure of buildings, the movement of goods constitutes traffic, there are rules and regulations governing warehousing operations, the workforce needs to have specific sets of skills, a communications system keeps tally and an auxiliary service would be the security system that all warehouses have. Any of the nodes in Figure 11.1 involves the six NETS shown in Figure 11.2.

Historically, every link in the supply chain operated independently and the transition of a node from being the destination of one link to the start of another required negotiation. The human communications involved were uncertain and at the international level compounded by language, cultural and political differences. Many links in supply chains are now functionally integrated and managed by a single controlling authority. This makes for smooth transition along the supply chain with the right information being passed at the right time to the right

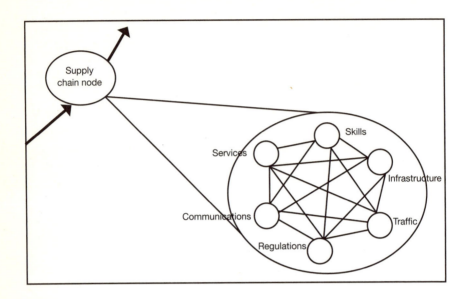

Figure 11.2   The six NETS

place and is increasingly supplemented by electronic systems for accounting, warehouse management, and transport logistics purposes. The trend is for fewer discrete control centres for these supply chains. Global commerce is becoming organized around competing value-adding supply chains (Robinson, 2002).

Supply chains, therefore, are a network of organizations that are cooperating through upstream and downstream linkages, with each organization adding value in order to be included.

Supply chain management requires communications directed at the first order of meaning in the form of logistic information and at the second order of meaning in the form of diplomacy, persuasion and courtesy to achieve coordination. Players in supply chains have to cooperate. This requires inter-organizational negotiation. In global supply chains it also requires inter-cultural communications.

Logistics is planning, assessing and coordinating the activities of each link in the supply chain. Since this is first order communication it can increasingly be done with the application of information systems. Logisticians will run models, paint alternative scenarios using both normative and simulation techniques, and apply sensitivity tests. This permits confirmation of physical capability and expected costs and risks. Supply chain managers, on the other hand, have to be able to

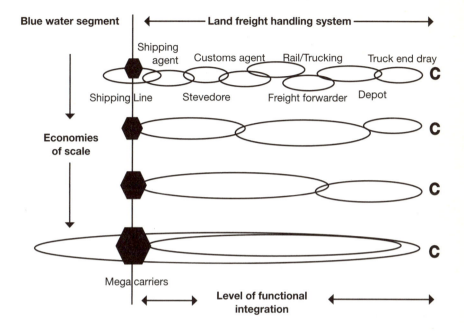

Figure 11.3 Elements in value-driven chain systems (after Robinson 2002)

Diagram courtesy of R Robinson.

evaluate alternative strategies and their political consequences and this involves politicking, diplomacy and negotiation.

Logisticians determine whether there is any point in moving products or people from one place to another. Goods deteriorate in transit, but upon arrival they need to retain sufficient value to be worth the effort of transportation. The value added by moving something from one place to another has to be sufficient for it to be competitive in the marketplace. The value of perishables disappears rapidly if they are not picked, packed and positioned elsewhere in a timely fashion. People moving from one place to another suffer "deterioration" with jet-lag, but they still choose to travel because they place more value on being at their destination than at their place of departure. It is the job of logisticians to assess the in-transit deterioration against the desired outcomes.

Communications within supply chain management

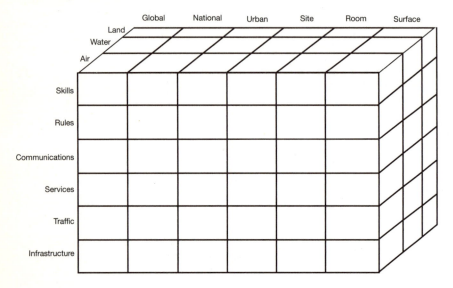

Figure 11.4 The supply chain matrix showing relationships between the three main modes of transport, the fractal levels of organization and the NETS

At every level of transport, in every mode of transport, there are the six different kinds of networks that enable transport which we have been referring to as the NETS. Figure 11.4 shows their relationship. It looks rather like a Rubik's cube, which is an interesting metaphor. Supply chain management seeks to align the faces between the cells to obtain the optimum sequence for channelling the flow of goods and people across multiple links.

Good communications within the management of supply chains involves the integration of technology, carriers, shippers, receivers and third parties. This integration cannot take place without communicating key information at the first level of meaning in a manner that is timely, capable of being understood by humans *and* computers and then being applied in an appropriate way. Quantitative first order communication about the conduct of supply chains increasingly becomes a function of information technology.

Methods of communication used in supply chains and logistics planning are continuously evolving. There are growing applications of m-commerce (mobile commerce) within the transport sector using wireless and cellular systems to direct and track vehicle fleets. More and more, the medium for communication associated with global logistics is direct electronic data transfer between computers using specific protocols. Electronic data interchange (EDI) is the process of communicating or exchanging standard business documents from one computer directly to another (Stock and Lambert, 2001). Connection between different business systems is greatly facilitated by the use of the internet (internet EDI) which is attractive because it provides public access to services via the world wide web (Turban *et al*, 2002). The search capability of web browsers helps potential trading partners find out about each other. They can also share data openly or in a restricted access manner once they are in an ongoing business relationship.

The management of global supply chains involves all components of transport from the different fractal levels from walking and horse and cart, to jumbo jet or container ship. These components are deployed to procure supplies, sort and package, assemble and store components, and distribute items to the next point in the supply chain. Each hand-off stage in the supply chain is a crossing point from one node to another in the supply chain matrix and a point where the risk of something going wrong increases. Prior knowledge and details of pending arrival help remove uncertainties. Feedback helps upstream participants learn how they might assist avoidance of congestion downstream. In essence, Figure 11.1 is the linear extraction of a pathway through the supply chain matrix shown in Figure 11.4.

The 'before transport' phase

Before transport takes place from any one node to another in the supply chain matrix there is a planning phase. This is the start of first order cybernetics. It is about determining the objectives of a specific transport project and deciding whether to undertake it. What is to be transported from where to where and when? Behind this is the question as to whether the NETS needed for the different levels and modes of transport are available and what they will cost. This then leads to the issue of what the overall costs will be.

To decide whether to go ahead with a trade deal, supply chain managers need to establish the relevant global distribution of available supplies, the location of potential or actual goods, assess the relative procurement or distribution costs for alternative means of transport within specific timeframes, assess risks, compile critical documentation and check legal and security requirements (Deloitte, 2004). Many individuals and businesses use the services of third parties to help acquire and analyse data as a prelude to decision making. Such third party logistics providers are discussed later in this chapter.

Decisions have to be made on what documentation is needed to initiate, transport and terminate a shipment, obtain adequate insurance, and effect payment. Some of these key pieces of information will be determined by specific regulations in force that will be encountered during the transport phase in different countries, states, towns, even depots. The selection of insurance providers may need comparative analysis. Some decisions will be taken on the basis of habit, and perhaps the level of confidence built between trading partners over a period of time. Clearly, someone selling something in a foreign market wants to be sure that they will be paid. Likewise, the buyer wants to be certain that the goods will actually be shipped, not diverted in transit, and will arrive in good condition at the agreed place at the agreed time, before they are willing to make payment. This necessitates involving a third party in whom both seller and buyer are confident. Such a service is provided by banks who deal in international currency and letters of credit transactions. There is a growing demand for banks to use electronic documents rather than physical paper transactions. Security for electronic transactions is improving with smart encryption systems which boost confidence in their use in e-commerce. There is a compelling requirement since the events of 11 September 2001 to meet export/import compliance and security regulations on a pre-clearance basis.

The 'in transport' phase

During transportation, first order cybernetics is at work. What is transported is steered towards its destination and any deviation from this objective has to be corrected. At the level of a single transport event circular causality may be under the control of whoever is in charge of a ship a plane or a truck At the level of supply chains however, cybernetic control shifts to a manager responsible for ensuring consignments or passengers are kept on course across the different links in the chain and this may mean changing whole links.

To ensure the integrity and security of cargoes, warehouse management systems (WMS) track goods using barcodes or, increasingly, RFID to track multiple products involving different carriers and to optimize storage locations. This is an information logistics function that will be increasingly automated. An example is the automated multi-level parking buildings for cars and air freight storage in specially designed facilities at Hong Kong's Chep Lap Kok international airport. Global air freight is being concentrated in a select number of storage warehouses at major freight hub airports. This follows from the improved technical capacity of jet aircraft to deliver from these centres to any part of the globe in less than 24 hours.

Voice communications are one way for key personnel involved in global supply chain management to apply second order communication. A single word like 'thanks' by its inclusion or omission and the tone or emphasis it is given can carry many meanings. Face-to-face meetings with their kinesics, proxemics and dress codes allow subtle cybernetic dynamics that resolve problems, allow power plays and initiate new ideas.

Hard copy of messages is still required in many businesses for audit purposes. The teleprinter (Telex) had a brief period of high use before being superseded by EDI messages and e-mails delivered over the internet to personal computers linked to printers. EDI provides a standardized way of conveying digital information that can be reformatted into common documents used by industry such as bills of lading.

Keeping a measure on stock levels is an area of data management that lends itself to electronic automation. As items are taken from warehouses, or checked out of stores, the relevant registers for those items are decremented. When stocks fall below predetermined levels, automatic replenishment ordering is set in train. Collection of data at point of sale leads to restocking of items that are selling quickly. The relevant information is conveyed automatically to the warehouse and delivery services. Conversely, slow moving stock might be discontinued and reverse logistics applied to their recovery and disposal. As customers fill their supermarket trolley it could be smart enough to keep track of items and display the size of the mounting bill. If customers change their minds about buying any item, they could dump it on a reverse logistics conveyor inside the shop. When their trolley passes through check-out, the customer's credit card will be charged automatically if it has been inserted in the trolley's reader and the correct PIN entered. The supermarket inventory is adjusted and if necessary reorders are initiated. Check-out staff sort out problems rather than deal with every transaction. Such procedures would eliminate checkout queues.

Bar codes are useful for tracking specific products into warehouses right through to specific storage locations, and then tracking those products as they are picked to order and delivery is made. If a bar code is to be used for tracking a product right through a supply chain, then every point of scanning at every fractal level needs to be able to interpret the message held within the bar code. This implies conforming to a particular standard. There are competing standards being used internationally. Logic suggests that industry should agree on a common standard in order to avoid re-labelling products.

RFID is a labelling system that can be read by radio. The label or tag is a microchip that carries information about whatever it is tagged to. The chip has an antenna so that it can transmit information to an RFID reader that has a transceiver to interrogate an RFID tag and read its information. This will include a unique serial number along with information about the product. Customs officials can ascertain the complete manifest of a container or delivery vehicle or vessel without having to stop that transport and physically check its cargo. All the items tagged can radio their identification codes on demand when passing surveillance points that are within the transmitting range of the devices. These RFID tags come in different forms. Since the tags travel with the goods and are not necessarily recycled, they need to be robust and cheap. The cost of producing these smart tags is being reduced as economies of scale become possible with widespread usage. There is also evidence (Deloitte, 2004) that users of these devices are obtaining financial benefits which outweigh the costs that inhibited adoption of the technology.

RFID technology is not affected by low light levels like the optical readers which scan bar codes. Tags on vehicles and containers can be read in foggy weather and at night and over greater distances than is possible with bar codes. RFID tags can be used to alert people about the condition of product such as perishable food inside containers. RFID seals on containers alert authorities if unauthorized entry takes place. It is possible to interfere with radio frequencies and so disrupt automatic identification, but the technology is becoming robust.

The introduction of RFID systems in transport and logistics could impact on the conduct of business and trade to an extent rivalling if not exceeding the impact of containerization in the past 20 years. RFID tags and reading devices allow precise identification of items and allow accurate capture of that information for subsequent processing. In effect, the invisible is made visible electronically. All the characteristics of items can be transmitted including time at position, security status, condition, and quantity. This enables tracking, monitoring, and warning of any irregularities. It also allows the accumulation of historical information that can be used for planning, review, and decisions regarding future shipments. Because they need a source of power, active tags which are used over longer distances than passive tags are more expensive. Use of nanotechnology in tag design will help reduce the amount of power required to activate information exchange over a range of distances from very proximal to many metres whether the items are standing still or mobile as is the case with cars passing through toll booths or under electronic gantries that control entry to particular areas at particular times. It becomes possible to think of every domestic animal and every manufactured item having an RFID tag that monitors its condition and whereabouts.

Those entrusted with guiding vessels, vehicles and aircraft need up-to-date access to the information databases relevant to their operations such as latest available maps or charts, weather conditions, and levels of congestion at terminals. Airline pilots carry briefcases crammed full of documents related to the routes they fly, and details of the airports they must negotiate from runway to terminal, and to a specific air-bridge or parking position. In the near future those cases need not be full of paper. Major airlines are now installing cockpit displays that will provide all the detail in a visual manner. Any modifications to procedures will be available instantly, no matter whether these derive from flight information services, air traffic control, or airport ground control.

Some airlines are also installing equipment for monitoring crew for their alertness, especially in long flights at night. Management and aircrew will determine the necessary work breaks, or whether extra crew will be needed to share the work load on the long flights that are now spanning half the globe such as between Singapore and New York.

Customers and regulatory authorities desire visibility in the supply chain from those who provide transport services and to know when their goods are likely to arrive. Carriers need to manage their traffic to extract optimum utilization. Regulatory authorities want to assess any threats to security and keep watch on trade. Increasingly, use is made of GPS to pinpoint where vessels, aircraft and land transport vehicles are located. One problem for security agencies who wish to delay the processing and release of consignments that are subject to investigation, is that the shippers of those goods can know through transparent tracking information that an unusual delay is taking place. That may give them sufficient time to avoid apprehension and interrogation.

The quest is still unfulfilled in this first decade of the 21st century to track an item globally between any two randomly chosen places in a seamless manner from the commencement of the journey right through to final delivery to the end user. That is because different tracking systems and standards have been developed within competing organizations and forms of transport. Universal tracking is technically feasible, but requires conformity in information capture and exchange.

The internet has helped enable different computers to talk with each other. The internet, or something like it, will be the means of achieving full tracking capability provided vital details are passed between the various in-house systems. Integrated internet-based tracking programmes 'deep-link' into the databases of carrier web sites to track individual container IDs (Bollen *et al*, 2004).

The 'after delivery' phase

First order cybernetics continue into this phase. Did the cargo or the passengers arrive, on time, intact? If not is there something that can be done about it? Proof of delivery is one method of confirming that goods were received at a specific location, at a specific time, and signed for by a specific person. If that information is captured electronically, including the signature of the recipient, it is inserted into the management database almost instantly via mobile access to the internet. When a customer phones to query a particular shipment's whereabouts, they can receive all the details including a fax back of the signature on the delivery note.

The main concern of collecting data in the after delivery phase, however, is with second order cybernetics. How well is the company doing? Does the data show that customers are satisfied? Do things arrive on time in good condition? How can the system as a whole be improved? This is not easy to do in supply chain management because a whole raft of different companies may be involved at different levels and in different modes.

Once items have been delivered or passed between intermediaries along supply chains, a receiving clerk or market analyst uses feedback information for accounting and billing purposes and to assess and benchmark how well systems are working. Records of movements and deliveries are used. Analysts interrogate these to implement inventory controls, to evaluate market penetration, and to conduct other aspects of market research. Managers within supply chains need the feedback information in order to know whether consignments were handled correctly.

Trace-back of items is not as simple as tracking. The task here is to dredge up the history of how an item or items came to be delivered from initial source through all fractal levels along the supply chain. Many providers of tracking capability do not keep historical information for long. They may keep the information archived for their own accounting and planning purposes, but it takes special requests to access that historical data. If a product is transformed or processed in some way during transit, this can pose an information break that is difficult to bridge.

Financial records help to evaluate the cybernetic process and are also a requirement for businesses and for taxation authorities. The automatic updating of transaction histories as goods move into and out of stores makes sure that customers are billed for the right amounts and disputes avoided.

Systems analysts can query transaction histories stored in linked databases often referred to as data warehouses. What takes place in one segment of the supply chain can be compared with what happens elsewhere. Data warehouses are a rich source of information to assist with forecasting future demand, and comparing alternative systems of supplier delivery. They are used to assess the 'what if' scenarios that can emerge in second order cybernetics. Key performance indicators used for benchmarking need measuring if they are to be useful. The question of whether the right information has been reaching the right decision makers at any point along supply chains is only answered if information systems are used to capture appropriate data. Data warehousing also allows evaluation of whether the operation is efficient and effective through examining the resources allocated to specific tasks.

Third party logistics

Contracting out (outsourcing) logistical functions to third parties is becoming common as third party logistics companies (3PLs) become established experts in assembling all the necessary information. This information becomes the basis for making informed decisions about the best mix of service providers and the best pathway through the supply chain matrix. 3PLs have the responsibility to determine the best route to follow and which agents will be engaged at various stages along the supply chain. 3PLs forge partnerships with suppliers and customers. They look to align their own and customers' information systems so that key information is captured and passed on free of errors.

Clients of 3PLs access information systems and technologies that enable them to specify the kinds of service they require and allow them to integrate their own operations effectively with a freight transportation system (Chow *et al*, 2000). Some 3PLs offer to take full responsibility for the complete end-to-end task on one bill of lading (BIL) covering inland transport as well as ocean voyages. Multimodal transport that involves an ocean segment is a candidate for 'through transport' on a single document.

Travel agents as 3PLs

A classic example of a supply chain manager is a travel agent. Travel agents combine their professional knowledge and experience to put together a plan that includes ticketing, insurance, tours, hotels and car rental. Traditionally they received a percentage of what they sold, but this function has changed with the use of web pages by passengers to work out their own travel plans. Increasingly transport services providers (TSPs) and ASPs are linked directly with travellers by the internet and using a travel agent is an extra cost for the customer. Knowledge 'bots' that help online customers find what they want improve with use and the internet now allows detailed previews of what potential customers will enjoy, including virtual walk-throughs.

Global containerized freight

Much of the world's freight is packaged in standardized containers that can readily be interchanged between different modes of transport and different vehicles. We can liken intermodal interchange of containers to the transfer between the three modal segments of the supply chain matrix. Pre-planning is needed to decide what type of container should be used depending upon the nature of the product to be shipped. There are different standard sizes: 20-foot, 40-foot and 40-foot high cube. Some cargo needs precise container temperature or atmosphere control. Other cargo is stacked in open containers. There are container types for liquids as well as dry goods. In some cases empty containers are repositioned before being repacked as the balance of trade at any given loading or unloading node is seldom even. Standardized containers have been an essential prerequisite for expanding the capability of intermodal transport. The inter-operability and ease of transfer of containers has greatly reduced the turn-around time at transport hubs and facilitated the development of supply chains.

Many global supply chains are highly complex, involving multiple partners and hand-offs associated with production and distribution, thus posing extended security risks. IT communications mean that an incident on one side of the globe is known on the other within minutes. This enables very quick response by firms to protect their brand image, possibly involving diversion into alternative markets or points of supply. There are various competing proprietary information processing systems that address supply chain event management (Coyle, Bardi, and Novack, 2006). A system such as Bolero (www. bolero.net) provides a seamless link between all participants in a global supply chain that offers a neutral secure platform enabling paperless trading between buyers, sellers, and their logistics service and bank partners.

Case study: the cool chain for meat

Any farming business is only viable (unless supported by subsidies) if there is a market for the commodities produced that earns sufficient return on investment. That presupposes that market intelligence about what potential customers want is conveyed to the farmer, perhaps by government agencies, farmer cooperatives, or downstream processing firms who purchase stock from the farmers. The farmer organizes production and arranges for all necessary inputs to be procured and delivered to the farm. That entails complex communications. In this case study we look at the production of cattle for the export meat trade from New Zealand to Europe. This is a longstanding trade involving a supply chain that stretches from one side of the globe to the other.

Timely communications

Meat-processing companies in New Zealand need to procure livestock from farmers in sufficient numbers and at the right time to ensure that they can fill specific orders for specific cuts of meat to be transported to northern hemisphere buyers in time for delivery when maximum financial returns are expected. This means farmers need to know when the processing companies want to receive their livestock. Stock and station agents are used by meat-processing companies to source livestock and arrange for their transport to the processing plant. Competing processing companies will announce their schedules and prices they are willing to pay. The outcome is that animals are transported in opposite directions over long distances between regions as meat-processing companies scramble to acquire livestock or farmers seek the best return on their production.

Agents need to give farmers sufficient notice that their animals are to be transported. Stock should be brought in from the paddocks and yarded to allow the animals to stand and empty their effluent. Stock should stand for around four hours before being transported. It allows the animals to calm. Stressed cattle with high pH factors do not convert into tender meat. If they are contaminated by their own effluent in the trucks while in transit, they are likely to become stressed and this stress is increased by the extra washing that must be performed at the meat-processing facilities. The importance of transporting unstressed stock for processing is illustrated in the following story.

> Knowing the negative impact of stress on quality, one specialist beef producer familiarizes his cattle with transport by using livestock trucks to move them short distances from one pasture to another on his property. The cattle come to expect a good feed after a short journey. They are docile in transit. The meat cuts from these cattle are premium grade and command high prices. Producers of premium meat can now establish a special supply chain relationship that runs from a particular animal in a particular paddock to a plate on a table in a restaurant that may be at the opposite side of the world.
>
> One day the farmer received a complaint from the end user's chef. Customers had complained about the tough beef. A particular consignment of this farmer's meat was not up to the high quality that they had come to expect. The cause was traced back to a particular relief driver of a livestock vehicle. This driver did not have skills in transporting livestock. Trying to catch up on

lost time he had subjected the livestock to a rough journey and they arrived stressed at the processing works.

Monitoring

Stowage of meat items into refrigerated containers (reefers) is a specialist task. Care must be taken to ensure that there is sufficient airflow around the container's contents to keep them at the right temperature. Before stowage, the goods should themselves be at the correct temperature. The cooling units are designed to maintain temperatures, not bring the temperature of the contents down to any required level. Temperature data loggers, when packed inside containers with the goods, chart temperature to prove, when the container is unpacked, that the journey was carried out without the container's temperatures going outside specifications. Many newer containers have built-in sensor equipment. These sensors sound alarms if temperatures threaten to move outside allowable limits. Sometimes they are monitored remotely if linked by radio frequency devices. Those in charge of the means of transport, such as logistics managers wherever they happen to be, can receive alerts by satellite communications.

It is necessary for the meat export companies to arrange for the transport of their products from New Zealand to foreign markets. Shipping space on vessels equipped to transport reefers is scheduled well in advance of loading. There are competing demands for these vessels. Contracts with shipping companies need to be agreed. The supply of containers to these vessels needs to be just in time to minimize temporary storage at container terminals where stacking space is at a premium and plug-in power points for cooling units limited.

Quality control

From meat packing to shipside can involve several means of land transport including rail and road or coastal transhipment from minor ports to the main ports of call for the international shipping lines. Coordination of movements requires effective communications and pre-planning.

Once on board a reefer vessel, the containers of meat cross the oceans and pass through regions of differing climate. Malfunctions can occur while the containers are in transit. Ships carry spare parts for containers of various makes. It is not unknown for the ship's engineer to effect temporary repairs on reefers, even using non-standard parts. However, it is not always easy to gain access to a particular box without resorting to drastic means such as cutting away railings, as one researcher found travelling 'jockey' with four containers on board a P&O Nedlloyd vessel (McColgan, 2001). Transhipment at intermediate ports in tropical climates is problematic. If reefer containers are left disconnected from power and exposed to direct sunlight for even short periods, internal temperatures are compromised. Adequate surveillance is necessary to ensure that this does not happen.

At port of discharge, the reverse process of unloading and transfer to land transport takes place with movement of the product into cool stores ready for final distribution to supermarket shelves. Quality control must continue all the way through this supply chain. It becomes apparent quickly if the product has not been properly transported as the shelf-life in the supermarket is greatly reduced.

Track and trace

Part of the process of selling a variety of meat exports into overseas markets involves packaging and labelling. Consumers are attracted by specific branding and origins of goods. Those who are marketing items seek to maximize the positive branding images. They are also quick to isolate their brands from suspect brands that are associated with outbreaks of disease.

Others look for reassurance that the beef products they are buying come from a disease-free source. Any allegation that a meat product from a particular source might in some way be contaminated has to be refuted successfully if the trade is to survive. This means that trace-back capability has to be built into the supply chain. Every individually packaged cut of meat displayed on the shelves of supermarkets has barcodes for initiating a trace-back process. Once the meat-processing company is identified, it may have to access the DNA specimens associated with the day of slaughter in order to trace meat back to individual farms. The same need applies to other food lines.

Data use

3PL firms record information about the sales of product and store it in relational databases to plan future marketing. The performance of the whole supply chain at any stage is reviewable. Poorly performing contracted parties are identifiable and can be replaced by new ones. It is possible to compare competing supply chains. Streamlining documentation avoids delays, reduces data errors and allows items to flow uninterrupted across national borders. With heightened security requirements for international container flows, including the need for advanced notification and certification of the contents of containers before they are loaded on ships, it is paramount that documentation complies with regulations.

Conclusions

Between 1950 and 1970 supply chains operated in an environment that has been classified as 'demand certain and supply uncertain' (Rahman and Findlay, 2003). There were relatively long product cycles and demand changed slowly. Uncertainty attached to the ability to supply. To combat that, inventory holdings were kept high, just in case product delivery failed.

The 1980s and 1990s saw greater capacity to control supply in conjunction with increased customization. The just-in-time model emerged as product lifecycles shortened and product choice widened. Firms contracted out their supply chain management services to third parties. Things improved. However, there is always some uncertainty in transport. The next chapter looks at this 'noise' and the problems transport has always faced. It also looks at new problems that have become apparent in transport environments in the new millennium with the terrorism event of 11 September 2001, the threat of pandemics and global warming. This has resulted in the use of less preferred supply chains, duplication of supply chains (second sourcing), higher levels of inventory, lower degrees of customization, investment in private security systems, and longer cycle times. Rather than simplification in supply chains, complexity is re-emerging.

The law of requisite variety applied to supply argues that each disturbance must be compensated by an appropriate counteraction (Ashby, 1991). This calls for management skills at the first order of cybernetics. But the epistemic changes discussed in the final chapter suggest a need for management that addresses the second order of cybernetics and looks to the future of transport at the corporate level and across supply chains whether they are local, national or global in extent.

12

Troubles in transport

It is important for the human race to spread out into space for the survival of the species. Life on earth is at the ever-increasing risk of being wiped out by a disaster, such as sudden global warming, nuclear war, a genetically engineered virus or other dangers we have not yet thought of. (*Online report of a press conference given by Stephen Hawking in Hong Kong, June 2006*)

Road deaths are a global epidemic on the scale of malaria and tuberculosis... Of the 1.2 million people killed and 50 million injured around the world in road traffic crashes, more than 85% of casualties are in low and middle income countries. Road deaths in these countries are forecast to almost double by 2020. (Lord Robertson of Port Ellen, speaking at the launch of a report for The Commission for Global Road Safety in London on 8 June 2006) [Online] http://www.fia.com/automotive/issue5/foundation/article1.html

Climate change is a far greater threat to the world than international terrorism. (Sir David King, Chief Scientific Adviser to the British Government, interviewed on BBC News (UK) on 9 January, 2004)

Then the whining school-boy, with his satchel,
And shining morning face, creeping like snail,
Unwillingly to school.

('The seven ages of man', *As You Like It*, William Shakespeare)

Introduction

Since the times of Shakespeare, children have dawdled on their way to school. Now anxious parents in developed countries drive their children to and from school and see them safely to the very doors of their classrooms. They worry about the traffic, kidnappers, drug peddlers, paedophiles, bullies and terrorists. There were three schoolchildren in the plane that ploughed into the Pentagon on 11 September 2001. We come back to where we began this book. The globalization of transport and communications brings a new dimension to the episteme we live in and it is not necessarily benign.

When we venture out into the world from a place of shelter we become vulnerable. In transport we are more at the mercy of the elements and rogue conditions in the environment. Failures in any one of the NETS will affect any transport initiative at any fractal level in any mode. When the infrastructure collapses and a road is down or a bridge is inaccessible, when traffic goes into gridlock, when people crash because they lack skills or fail to follow regulations, when there is fog and drivers cannot see the traffic signals and emergency services cannot respond because they are overstretched, there is trouble in transport. Transport systems have always had to allow for such problems and one of the roles of auxiliary services NETS is providing for the safety and security of people and goods. However, since the events of 11 September 2001, there has been a heightened concern with transport security that has caused a decline in transport services. Travel insurance rates increase and additional security measures reduce the pleasure and convenience of travel and transport.

This chapter seeks to put a perspective on the problems that transport systems face now and in the future and on how IT is likely to be involved in their solution.

Terrorism

The UN high-level panel on threats, challenges and change defines terrorism as any action intended to cause death or serious bodily harm to civilians or non-combatants with the purpose of intimidating a population or compelling a government or international organization to do, or abstain from, any act (Davidsson, 2005).

Acts of terrorism are acts of communication and like all communication have two orders of signification. There are the acts of terrorism themselves. The mayhem and carnage they cause and the numbers killed and the damage involved is first order communications. Second order communications come in the feelings of fear and confusion that the threat of terrorism brings. This extends beyond the place and time of a particular act of terror so that the threat of further acts is sufficient to produce terror. Mass media have globalized terrorist threats by showing the world acts of bizarre violence in graphic and convincing detail. The bigger and more horrific the act of terrorism, the more news cover it gets and the more terror it inspires.

Terrorists use transport and target transport. Pedestrians with suitcases or body belts may carry explosives or drive cars and trucks loaded with explosives to sites where the consequences of detonating them are horrific. Such sites are often crowded streets or transport vehicles such as buses, trains and planes. As a result, transport systems are heavily policed and their cargoes and passengers become suspect.

According to the US Department of State (State Department, 2006), in 2005 there were 11,111 incidents of terrorism worldwide. The consequences included 14,062 people killed, 24,705 injured and 34,780 kidnapped (Perl, 2006). How does this compare with the other threats transport has to cope with?

War and genocide

Terrorism between states manifests itself as war and within states as arbitrary arrest, torture, suppression of freedoms, imprisonment without due law, the disappearance of dissidents and ethnic cleansing.

The First World War killed some 15 million people, half of whom were in armed forces. The Second World War killed some 55 million, the majority of whom were civilians (White, 2005). This is matched by the level of violence in intra-state terrorism. At least 5 million Jews died in Hitler's concentration camps. Estimates of the numbers who were killed by the state in Stalin's Russia vary from 9 to 50 million. Estimates for the labour camps, purges and cultural revolutions of Mao Zedong's China are in the order of 40 million. Millions have been massacred in Africa in Rwanda, Darfur and the Congo. Mass graves are still being discovered in the Balkans. The 'disappearances' of thousands of people in Argentina, Chile and El Salvador continue to haunt Latin America.

Acts of war are acts of communication similar to terrorism. The means of communication are transport systems. Tanks, warplanes and battleships carry bullets, bombs, shells and combatants towards the opposing side so that at one level they can kill people and cause mayhem and at another level break an enemy's will to fight. Few things fit Shannon's model of communications as well as a soldier as source firing a gun that transmits a message in the form of a bullet that is received by the body of an enemy who is its target destination. Intercontinental missiles are more effective than arrows and tanks more efficient than horses, but the basic communications principal of war has remained the same throughout history.

The opposing sides in a war seek to destroy each other's means of transport. To do this they try to improve their own means of transport while wreaking havoc on that of the enemy. Wars can speed advances in transport technology. The development of rocketry and drogues by today's defence services suggests the way of the future for air transport.

Invasive wars such as those in Iraq, Afghanistan and the Falklands cause severe damage to the transport infrastructures of the areas fought over. To global transport systems they present temporary no-go zones. However, a regional or global war would have serious consequences for all countries, especially those with a limited range of economic activities. As globalization makes countries economically interdependent, they become increasingly at risk from any widespread collapse of global transport infrastructures because they have lost self-sufficiency.

A characteristic of countries that have nuclear weapons is that they behave belligerently and countries that are belligerent seek nuclear weapons. The threat of a global nuclear war receded with the end of the Cold War, but was not eliminated. With nuclear proliferation it will return. Even if such a war was contained within one section of the globe it would profoundly impact on all transport systems that have adopted IT. The Star Wars concept of nuclear war shifted the focus to fighting in space. High altitude nuclear explosions produce an electromagnetic pulse (EMP) that extends far beyond the location of the explosion. While it may have little physical impact on people and buildings at a distance from the combat zone, it can knock out sensitive electrical apparatus. Transport and communications systems become increasingly dependent on microprocessors that would be vulnerable to EMP. Aeroplane engines could stop and their communications systems fail while they were in flight. Telecommunications and computing systems could collapse. The more societies become dependent on IT, the more vulnerable they become to nuclear war. Critical telecommunications systems could be hardwired to protect them from EMP, but the privatization of telecommunications services means that national interests in survival have been supplanted by global commercial concerns.

Threats from the natural environment

Ships and aeroplanes move through the volatile mediums of air and water and are subject to its extremes. Railways and roads get blocked by snow, avalanches and floods; earthquakes rip them apart and bridges are washed away. Over time, the design of vehicles and transport infrastructures improves to deal with such natural threats. Ships and aeroplanes are made bigger and stronger and more able to withstand all but the most freakish natural events. But freakish natural events are becoming the norm. The tsunami of 2004 killed over 300,000 people, including many travellers. In the following year the Pakistan earthquake killed over 80,000. The number of deaths attributable to Hurricane Katrina did not exceed 2,000, but the damage to property and people's lives was enormous and draws attention to what is likely to be one of the great issues of this century. With global warming and rising sea levels, for how much longer will the coastal lowlands that harbour so much of the world's population be habitable?

It is in the nature of science to be critical and to recognize its own limitations. It is hardly surprising, then, that there is so much contention over the issues that surround global warming. There is scientific consensus that the earth has been getting warmer since the Industrial Revolution and this can be correlated with the presence in the atmosphere of greenhouse gases, the most critical of which is carbon dioxide. What is in contention is the future projection of this trend and its implications (IPCC, 2001a, 2001b). Is the rate of warming increasing? Are we moving towards a trigger point where climatic change is of such an order that, as Stephen Hawking speculates, it could threaten the very existence of our species on this planet?

The Victorian scientist John Tyndall discovered that carbon dioxide could absorb heat radiation from the earth and in consequence that it was variations in carbon dioxide levels that brought about the various ice ages. Taking this as his starting point, Sir David King, Chief Scientific Adviser to the British Government, set out the issues of global warming in the ninth Zuckerman Lecture (King, 2002). The cyclical occurrence of the ice ages can be measured from ice cores from

Antarctica and studies of air bubbles trapped in ancient ice show the associated levels of carbon dioxide. Through the last ice age these were in the order of 200 parts per million (ppm). Between 17,000 and 12,000 years ago, the level rose to about 270 ppm and this rise marked the end of the last ice age, the beginning of our current climatic period and human history. Today we are at 372 ppm and still rising.

Extrapolating from such data, global temperatures will rise between 1.5 and 5 degrees centigrade by 2100. By then the Arctic sea ice will have melted and sea levels have risen between 0.5 metre and 1 metre. The Antarctic, with about 90 per cent of the global ice mass, will take longer to melt, but by the time it does, sea levels will have risen an extraordinary 100 metres.

Global warming also means there will be more energy in the atmosphere and so more extreme and violent weather conditions. The images of droughts, violent storms and floods we see on television are not there because there is a lack of news. They constitute the news. Events that used to occur every 100 years become everyday events. Transport vehicles and infrastructures are made to have lifetimes of 20 years and more. They were not built to withstand the new extremes. Transport engineers find that they need to reassess standards and imagine scenarios that were previously unimaginable.

When bad weather is forecast, flights are cancelled, ships stay in port, schedules are adjusted and people are advised not to travel. Planes in transit fly around or above turbulence and ships change course to avoid cyclones and hurricanes. When violent weather is moving towards a location, the people in it are advised to move to a safer location or take precautions. Much of the danger from freak conditions comes from a lack of information, failure to communicate what information there is to where it is needed and a failure to act on it. Many of the people killed by the 2004 tsunami could have survived if they had had enough warning to get off the beaches and head for high ground. There was no tsunami warning system in place for the Indian Ocean as there is for the Pacific.

People piloting ships and aeroplanes in the first half of the 20th century worked out the weather from what they could see in the sky, feel in the wind and intuit from a barometer. They only knew the conditions at the place they were in. IT has changed this. Now individuals have access to up-to-the-minute weather reports from the internet, portable radios and mobile phones. Computers have improved weather prediction. Telemetric systems become ubiquitous. Our ability to sense conditions globally from the perspective of satellites down to the details of microenvironments improves.

Even when warnings are provided, however, people still drive at speed in fogs, still walk in mountains without letting anyone know what they are doing or checking the weather, and still go out to sea without life jackets and radios. Few people know what to do in a crash or have basic first aid or survival skills. A major danger in transport arises from ignorance. One day we might be safe inside nanosuits that monitor environmental conditions and advise us what to do in danger, but until that day it is a matter of concern that educational systems pay so little regard to teaching basic transport skills when they are such a major part of people's lives.

Danger from disease

A pandemic is an epidemic spread globally by transport systems. The biggest danger humanity has ever faced was the Spanish flu pandemic of 1918. It derived its name from the fact that, unlike the countries that were involved in fighting the First World War, neutral Spain did not have censorship and so was the first country to report the seriousness of the disease in full. It is estimated that a third to a half of the world's population was affected. Approximately 25 million people died in two years. We now know that it was a form of avian flu and expert pandemicists believe that there will inevitably be another flu pandemic and that it could be on the scale of the 1918 Spanish flu outbreak.

Transport systems transmit disease between people, and between animals and plants. The bubonic plague that arrived in Europe in 1348 killed approximately 20 million people in six years. It was reputed to have been brought by ships' rats as a consequence of opening up new trade routes to the East. Cholera spread from India in the 19th century to China, Europe, the United States and Russia. In turn, the voyages of discovery out of Europe introduced whooping cough, measles and smallpox with deadly effect to communities with no immunity to them. Typhus is known as ship fever and camp fever because of the way it spreads. It is associated with war and catastrophe and the march of armies. AIDS is a slow-moving pandemic that is not yet under control. There are diseases with the potential to become pandemic, such as Lassa fever, Rift Valley fever, Marburg virus, and Ebola virus, which are still geographically restricted in their occurrence. Antibiotic-resistant superbugs may cause the diseases we regard as curable to take on a new malignancy.

When we improve transport systems we facilitate pandemics. In the past when global travel was by sea, the symptoms of diseases that had been contracted before travel would appear during a voyage. Ship's companies and passengers were inspected before they landed for signs of disease and if there were any, the ships would have to fly the notorious 'yellow Jack' flag, which meant they were in quarantine and not allowed contact with the land. Quarantining is the single most effective strategy for stopping pandemics, but air travel is now so fast that if symptoms of a disease were not evident on departure, there is little time for them to emerge in flight. Someone can unwittingly carry a disease from one side of the world to another within a day.

In 2003 the WHO alerted the world to the appearance of a severe respiratory illness of undetermined cause that was rapidly spreading among hospital staff in Vietnam and Hong Kong. It became apparent that the new disease was also travelling along major airline routes. Within a month of the first alert, some 3,000 probable cases and more than 100 deaths had been reported from more than 20 countries on all continents. The world held its breath. Yet within another month the disease had been stopped in its tracks. The WHO attributed this to a communications strategy using IT which emphasized rapid and open reporting of cases and international collaboration so that the traditional control tools of isolation, contact tracing and quarantine could break the chain of transmission. However, the SARS outbreak also highlighted major weaknesses in responding to a global health threat. No country had the surge capacity to cope with the SARS caseload, without dangerously extending existing health services, especially since health-care workers – the frontline troops – were themselves frequent victims of the disease (Brundtland, 2003).

Global concern started in 2005 that an avian flu pandemic on the scale of the Spanish flu of 1918 was imminent. Countries began to draw up contingency plans. These began with compulsory quarantining of countries themselves that would involve a cessation of global transport. This would be followed by quarantining cities and regions with road blocks that allowed only essential services through. Interurban transport would be limited. It might even become necessary to close shops, schools, hospitals and work places. In such a scenario the internet, which originated as a communications network that could survive an atomic attack, would come into its own. Teleservices such as teleworking, telemedicine and tele-education would become critical because they used telecommunications rather than transport.

Transport has always had to cope with problems from disease, natural disasters and the violence people are capable of. We turn now to the problems of land, air and water transport that are peculiar to the times we live in.

Automania

Global warming is primarily caused by carbon dioxide emissions. The biggest source of carbon dioxide emissions is transport and the biggest source within transport is automobiles. Automobiles are now the dominant means of terrestrial transport. They have made possible the development of the giant conurbations in which most of the world's population now live and the traffic congestion from which they suffer. They are a major cause of air pollution and they kill. The annual death toll in the United States alone from automobile accidents is in the order of 40,000. This number far exceeds the total number of deaths around the world from acts of terrorism. Yet no one declares war on the automobile.

The second order of communication in car advertising is often linked to speed and adventure and may include such phrases as: 'With this little beauty you can go like the wind', 'What is life for if you don't take risks?' Every country has problems with speeding yet none take such simple steps as refusing to license any car that has the capacity to go above the speed limit or making car advertisements carry a health warning, like a packet of cigarettes.

The conurbations where most of the world's population lives become car-dependent. People take their car to get to work, to shop, to go to the park, to the gym, to the church or to take the kids to school. Each of these journeys involves a mini-gridlock as everyone arrives at the same time and looks for somewhere to park. Not only do these multiple journeys mean that urbanites spend a lot of time on roads, but when they are not in traffic they are using roads for parking.

Every year the number of cars in the world increases. With current trends the number of cars in China and India will rival those in the United States within the next quarter of a century.

Cattle class

People who own planes drive to an airstrip and arrange by telephone to have their plane waiting to take off directly. At the national level of flying there is no immigration or customs to clear and it is possible to check-in up to half an hour before a flight. It is in the large international hub airports that air transport is in decline in the quality of service it provides.

In part, this is a consequence of the way air traffic has been doubling every decade since the 1960s. Big airports compete for space with the metropolises they serve and have become part of their gridlock problems. The main mode of access to airports is by car and the number of cars on the roads going to airports has expanded to match the growth in air traffic. The growth of air traffic is increasingly determined, not by the provision of new runways and terminal infrastructure at airports, but by the provision of public transport alternatives that alleviate the congestion that comes from the use of cars to access airports (Humphreys and Ison, 2005). Most people using a hub airport will spend over an hour getting to it from downtown. International passengers are required to check in two or three hours before take-off time. They then face a further hour when they arrive in their destination country going through customs and immigration, followed by another hour or more in traffic to their final destination. If they have to make a change to a domestic flight or another international flight they will spend another two hours in transit. The time passengers actually spend in flight steadily decreases but the time they spend on the ground steadily increases.

The actual time involved in interacting with essential airport services, as when passengers actually check in, interact with customs and immigration officials and pass through security checks, is in total less than 15 minutes unless there is a problem. This time can be reduced if these functions are automated. Automatic check-in could be done at the kerbside so that passengers would enter an airport with only their hand luggage. The United States has mandated the use of RFID-tagged passports with biometric information security checks from 2006. Malaysia already has them in use and many countries plan to develop such systems. These e-passports will carry digitized information on their owner's face and fingerprints. They should reduce fraud, improve authentification, eliminate repeated filling in of immigration forms and speed up passport checks (Juels, Molnar and Wagner, 2005).

The 15 minutes spent in essential airport interactions could be reduced to 5–10 minutes, but the real problem is in the weary hours people spend queuing to get these services. More processing staff and more automated processing units would reduce queues and the peaks and troughs of passenger flows could be smoothed out if, instead of

Figure 12.1 Changi airport automatic check-in system

Trial at Changi airport of an automatic fast-track check-in system that uses handprints and face scans

telling all passengers for a flight to arrive at the same time, they could be asked when they would like to arrive. This information could be used to organize estimated times of arrival for each passenger.

If airports used IT to track and organize the flow of passenger traffic through an airport as they do with flight traffic, the time spent in international airports could be greatly reduced and planes could be trickle-loaded. However, the main source of revenue for large international hub airports is as likely to be from the concessions that use it as from the airlines. Restaurants, bars, newsagents and duty-free outlets provide auxiliary services for travellers, many of whom enjoy doing some shopping while they are travelling. However, there is a fundamental opposition of purposes between passengers who want to complete their journey as quickly as possible and concessionaires who want to have them in an airport for as long as possible spending money.

International airports have become shopping malls that passengers have to pass through to get to their planes.

Air travel is a safe form of transport. It also used to be a very pleasurable way to travel, but not any more, at least not for the majority of travellers in economy class. Seating space has been reduced. Toilet facilities have become inadequate. Food services have declined. The dangers to health of sitting in cramped space become apparent. The recycled air in a plane usually consists of 50 per cent fresh and 50 per cent stale air. Bacteria and viruses that cause illnesses like colds, flu and pneumonia become airborne when passengers talk, cough and sneeze and then are circulated. (See the AviationHealth website: http://www.aviation-health.org/news/browse.php?load=Infectious.html.)

Seats face the wrong way for safety. In the event of a crash, the waist belts passengers are instructed to wear would bear all the momentum as passengers were jerked forward by the impact. Jack-knifing on waist belts causes serious damage to the spine and internal organs and the effect in an air crash is likely to be every bit as serious as in a car where, precisely because of this effect, shoulder harnesses have long been compulsory. If seats faced towards the rear of airplanes, people would be pressed back into their seats on impact and survival rates would be much higher. For this reason on take-off and landing cabin crew sit facing the rear wearing shoulder harnesses.

The security measures that have been put in place since the events of 11 September 2001 have added to the growing problems of international economy air travel. These measures can be unobtrusive. Before a flight takes off, its passenger list is lodged electronically with the receiving country's authorities for analysis. All items stowed in the belly cargo bays of aircraft flying internationally are screened for content. Every aspect of an international airport is under continuous camera surveillance. In the SARS epidemic of 2003, Changi airport introduced technology that could remotely sense body temperature and unobtrusively monitor passengers for fever as they filed past check points. Suspect cases were detoured for more complete checks and questioning.

Passengers are every bit as concerned with security as the operators of a transport system. Safety explanations are accepted with good grace and people patiently accept the need to pass through metal detectors. However, this good will on the part of travellers is being eroded by heavy-handed security systems. The increase in security checks allied to the decline in the quality of services generates friction. Passengers become more prone to argue and complain than in the past. The response from security services is to call for harsher measures to be taken against angry passengers. In a submission to the Joint Committee of Public Accounts and Audit Review of Aviation Security in Australia, the Australian Services Union commented 'The reasons for air rage are complex, but include... the failure of consumer expectations of air travel to coincide with reality and the vagaries of air travel including delays, overbooking, flight cancellations and baggage limitations... as security at airports is tightened air rage will increase and it must be met with sanctions against the offenders.' (Quoted by Steve Creedy in *The Australian*, 19–20 October 2003.)

Figure 12.2   The interior of a full aeroplane

Imagine what this place would be like if the plane crashed and a full load of people were all trying to get out of their seats to emergency exits in a dark smoke-filled cabin. Airlines have made it even more difficult to get in or out of a seat by reducing seating pitch at a time when people are increasing in girth.

It was passengers who stopped the terrorists in the fourth plane on 11 September 2001. Instead of treating passengers as potential terrorists, they could be enlisted in the prevention of terrorism. If passengers knew what procedures were in place to deal with a terrorist event, how they could help and what they should do, they would be more likely to cooperate and less likely to panic.

Boat people

In 2004 there were some 17 million refugees, mainly in the developing world, moving from one poor country to another on foot to flee famine, war or genocide. Many refugees return to their native country when the troubles they fled from end. The movement of refugees by sea to developed countries is not on the scale of the mass migrations from Europe to the Americas in the 18th and 19th centuries, but this may not continue to be the case. Many of the poorest people live on or close to water in areas vulnerable to rising sea levels and the violent storms that accompany global warming and boats are the transport vehicles they know about. In a special report on natural disasters entitled 'South Asia floods claim 1,100 lives' published in *The Guardian* (London), 28 July 2004, Ahmed reported that the 2004 monsoon floods in Bangladesh alone engulfed two-thirds of the country and affected more than 25 million people.

The Office of the United Nations High Commissioner for Refugees (UNHCR) estimates that 40 per cent of refugees are less than 18 years of age. Managed migration and integration of refugees could be a means to rejuvenate the ageing populations of the developed world. Instead, developed countries adopt a fortress mentality in which refugees are coming to be seen as a new class of criminals and refugee camps as a breeding ground for terrorism.

The perfect storm

There are other dangers that face transport (Posner, 2004). Trains and trucks that carry radioactive waste, explosives and dangerous chemicals can crash. Cargoes can harbour insects, seeds, microbes and viruses. Oil spills cause massive ecological damage. Trains and gas stations are robbed. Lonely unlit roads are dangerous. Piracy is as difficult to stamp out today as it was when the Phoenicians tried. Smuggling is as old as customs duty.

The concept of the perfect storm comes from a hurricane that struck the North Atlantic in 1991 with extraordinary force because of a conjunction of weather events, which by themselves would have been normal events. Hurricane Katrina was already a meteorologically perfect storm when it hit New Orleans and the Mississippi Coast in 2005. Such a storm in itself is a major threat to transport. In this case, it was compounded with problems in the networks that enable transport. Transport infrastructures were inadequate. Transport services were lacking. Communications collapsed. Law and order broke down. A million people trying to evacuate the area were gridlocked. As a result Hurricane Katrina became the costliest and deadliest storm in US history.

Compared to war, genocide, pestilence and natural disasters the kind of terrorism the United States has declared war on is a relatively minor threat to transport, but it is one that grows. It has the capability of combining with other threats to transport to create a perfect storm of nightmare proportions. Richardson (2004) draws attention to what could happen to transport and the whole world if ships rather than aeroplanes were used by terrorists to destroy cities rather than buildings.

13

Scenarios of globalization

The future is there... looking back at us. (Gibson, 2003: 57)

Introduction

We come back to where we started. Improvements in transport improve communications and improvements in communications improve transport. Positive feedback loops lead to systemic growth that in a finite system ultimately causes chaos. Humans have used the interaction of transport and communications to extend their activities across space to the point where they can now organize and act at a global level. But the earth is a finite system.

The totality of the paradigm shifts in transport and communications in the 1860s constituted the epistemic shift known as the Industrial Revolution, in which a scramble for empire took place. The 1960s saw yet another set of mutually reinforcing paradigm shifts in transport and communications that amounted to the epistemic shift we call the Information Revolution. However, there are no more territories to take and no more vast resources waiting to be tapped and the forces of chaos begin to apply. The problems emerging in transport described in the previous chapter are indicators of coming chaos. They constitute negative feedback that advises the need to change what we are doing in transport. We turn now to some of the alternatives in the form of futures scenarios that would require a paradigm shift for it to be implemented. These scenarios are based on the technologies discussed in previous chapters. They do not take into account paradigm shifts in biotechnology, the rapid growth of the global population, the ageing of populations in developed countries, the way women increasingly dominate education, and the economic rise of China, India, Russia and Brazil.

The doomsday scenario

Down the ages we have lived by the assumption that the more we improve the means of transport and communications, the better the human condition. And so, down the ages we have seen people and the things they produce and consume transported at ever-increasing speeds in ever-increasing quantities at every level of transport above that of the pedestrian. We believe this is progress. We put our faith in the advice of Adam Smith that the wealth of nations lies in the maximization of free trade. The world's richest countries maintain their standards of living by endlessly improving their transport systems and developing countries are developing because they are developing their transport systems. The peoples of China, India, Latin America and South East Asia who have been filling the ships of the world for the last 200 years now aspire to own a car.

Can the world maintain the present level of transport activity, let alone support its exponential expansion? Transport has always had to face the dangers that come from war, thieves, disease and bad weather, but there now comes a new threat of a different dimension. Over the last 150 years we have developed an economic system that depends on the global interchange of goods. The transport and manufacturing industries that make this possible in turn depend on cheap energy from coal and oil. The gases released into the atmosphere by the combustion of these fuels, in particular carbon dioxide (CO_2), cause global warming. The consequences, in the form of melting ice caps, rising sea levels, violent weather, climatic change and the flooding of flatlands will, if we continue as we are, prove catastrophic for the human race. The problem is not that the amount of oil in the world is peaking, but that it did not do so long ago. If we carry on doing what we are doing in transport, we invoke the doomsday scenario.

The low carbon economy scenario

A straightforward response to global warming is to reduce the emission of carbon dioxide. It can be done. Agricultural and industrial practices can be changed to sequester instead of releasing CO_2. Transport can use alternative energy sources. The problem is that the expansion of the world's economy since the 1960s has been based on transport and electricity that use oil as their primary source of energy because it is cheap and convenient. So much so, that goods produced with the aid of oil-powered equipment in developed countries can compete with equivalent products produced with peasant labour at the other side of the world.

As yet there is no other form of energy as cheap as oil. The infrastructure that makes oil so readily available is in place, the tankers and pipelines that deliver oil are there, the regulatory apparatus and communications systems and the skills that are needed have evolved. The investment in the continuance of dependence on oil is massive. Developed countries would lose their economic advantage without it. Developing economies are striving to match them. The will to kick the oil habit is not yet manifest. The world waits for the weather to get worse and seeks interim supplementary solutions for its power needs in the form of hydrogen, wind, tidal and solar energy, carbon capture and storage and by revisiting nuclear power.

Carbon taxes are another interim mechanism. They are an outcome of the Kyoto Protocol of 2004 when the industrialized countries, with the exception of Australia and the United States, committed to reduce their carbon emissions. The proposal is that nations that produce more CO_2 than they absorb should purchase 'absorption ability' from nations that produce less CO_2 than they absorb. Carbon credits are the currency of this new economy. They have a face value of one tonne of CO_2. Their actual exchange value in monetary terms is market-dependent and likely to be very volatile. The assessment procedures for the absorption and emission of CO_2 are fraught with uncertainties. Whether carbon credits will have an impact or provide a new field of fraud remains to be seen.

The u-scenario

The prefix 'e' as in e-commerce, e-learning and e-books is used to denote the adaptation to the internet of a traditional form of communication.

In Japan it is now being replaced by the prefix 'u', as in 'u-Japan' instead of 'e-Japan', to describe the next stage in Japan's evolution as an information society. The term u-Japan was promulgated by the Japanese Ministry of Internal Affairs and Communications in December 2004 as a goal for 2010. The scenario involved is the universal adoption of ubiquitous IT in the sense of miniature tags, sensors and robots enmeshed into the general environment so that everyone would have anywhere anytime access to broadband IT. It follows on from the term e-Japan which had as its aim the establishment of an advanced broadband infrastructure by 2005. (See the website of the Ministry of Internal Affairs and Communications on Information and Communications Policy: http://www.soumu.go.jp/joho_tsusin/eng/index.html.) It is the kind of scenario that can be envisioned as the technologies described in Chapter 6 become commonplace and widespread; an environment where broadband IT, informed by AI and accessible through clever clothes in HyperRealities, becomes possible through nanotechnology. IT appliances such as RFID tags and telesensor and telemetric devices will be embedded in every facet of the infrastructures we use and the appliances we handle and, in all probability, in ourselves. We are shrinking computers and the devices attached to them. From being a handful of huge room-size devices that only the richest countries and corporations could afford, computers have become handheld devices that are manufactured in their millions. But they will continue to shrink until they are nano-devices invisible to the naked eye and manufactured in billions. It is this that makes them ubiquitous and radically changes the environment. It is this that makes possible pilotless planes, autonomous automated cars, clever clothes and transport systems that manage themselves.

The adoption of nanotechnology will profoundly impact on the bulk transport of raw materials and finished goods. What is manufactured will be done where it is wanted. There will be no need for transport systems that bring coal and iron ore together or carry finished goods across the world to markets. The use of bulk carriers by land and sea will decline. Instead, small quantities of valuable and perhaps volatile nanotechnology stockfeed will be transported by air. The massive terminal infrastructures that have been developed to handle very large ships and their cargoes could go the way the old wharfsides and warehouses went with containerization. The distributive systems of manufactured goods will see a reduction in volume as shops and those parts of supermarkets that sell manufactured goods become increasingly redundant. The distribution systems of trucks and vans that is putting growing pressures on urban and national roadway systems will decline. Freight transport will shift from moving very large quantities long distances to the widespread distribution of small quantities over short distances.

Nanotechnology may reduce the volume of freight transported, but it cannot reduce the size of people and transporting people is the biggest source of income and the biggest pollutant in transport. However, automobiles and aeroplanes built with nanotechnology would be several orders lighter, stronger and more fuel-efficient than they are now. Planes would be able to fly higher and faster and, while they could become bigger, they could also become smaller, safer and more difficult to shoot down. Applied to rocketry, nanotechnology could be the factor that realizes the expansion of transport into space.

Smart villages and clever cottages

Logic would suggest that improving telecommunications will reduce the need for transporting people, but this has not happened yet. Telecommunications may save on specific acts of transport, but seems to stimulate transport systems in compensation. The telephone call that saves one journey is just as likely to lead to another. However, telecommunications are still in transition. We are shifting from dumb, analogue, narrowband, copper wire, place-to-place telecommunications, to smart, digital, broadband, fibre optic, person-to-person, cellular and computer-mediated telecommunications. It is the equivalent in transport of shifting from using footpaths to using motorways. So far, the changes we see are in the growing use of mobile phones, faster internet access and the ability to download music and videos, but the new infrastructures are not yet fully in place. When broadband internet, clever clothes, HyperReality and embedded AI become commonplace, the emphasis in our interactions with the world and with each other could shift from transport to telecommunications.

As it becomes easier to go somewhere as a telepresence than to go there in person, when we can make the things we want at home precisely as we want them, when we can bring the world to our door instead of having to go out and find it, then we may see a reversal of the transport patterns of the industrial society. It is less than two centuries since people began leaving the countryside to work in the shops and factories of cities, and the development of railways, metros and bus and tram services saw the growth of commuting and conurbations.

When the dominant way of life was a rural one and people lived in villages, they grew much of the food they needed, kept a few animals and depended for most of what they needed on what they produced themselves or could trade with their neighbours. They walked to market. Manufacturing took place in the worker's home. Pack animals brought raw materials to people's houses and took away finished goods. It was called cottage industry.

Will we see a reversion to such a way of life as people set up nano-factories in 'clever cottages'? There is a move back to the country led by professionals involved in the new technologies. Life-stylers with a Valhalla complex buy remote sites in areas of great scenic beauty that

Figure 13.1　The old packhorse bridge and 19th-century bridge

The old packhorse bridge in the foreground was the only way across the river until the railway bridge behind it was built in the 19th century. The narrow apex of the bridge gives some idea of the nature and volume of traffic in pre-industrial times. The railway transported fish to the markets of industrial towns and holidaymakers to coastal beaches. It was the main means of transport in this area until the volume of motorized traffic made it necessary to build the road bridge (to the right of the old bridge). This is now the main means of transport. Some of the people living in this rural area are teleworkers who go into town by car only when necessary.

are not easily accessible. They seek a return to an idyllic rustic setting and have a vision of a greener, more environmentally friendly way of living made possible by advances in IT. These teleworkers have their children attend virtual classes in virtual schools. They go teleshopping and telebanking and if they are ill call on telemedicine.

These people are not the rural poor who made up the workers in the cottage industry of yore. They are a new elite of teleprofessionals concerned with the preservation of the scenic sites they have newly colonized. They live in clusters of lodges, socialize at golf clubs, travel to their rural residences in SUVs and private planes and push up property values so that the people who originally populated the remote scenic sites they favour are forced out or become a new servant class.

This is where the new rich are going, but where will the rest of the extra 3–4 billion people live as the world's population grows from 6 to 9 billion in the next 50 years? Population densities in rural areas may increase, but so too will the giant conurbations.

Singapore, as a small island state with no hinterland for expansion, has pioneered the idea of smart offices and smart homes in smart buildings in smart blocks in smart cities in a smart country. By 'smart' is meant the using of IT to make the environment in which people live and work more energy-efficient, more secure and more integrated so that everything is to hand and the need for motorized transport greatly reduced. Buildings, rooms and elevators can recognize individuals and allow them entrance and adjust the climates to match the occupants' preferences. The trend to compact the use of space is encouraged by making cars and their use of city roads expensive and the cost of using telecommunications cheap. Singapore is a city where people walk a lot. They also spend a lot of time in elevators and on escalators.

The matrix

Most of the world's population lives in cities. Growing pressure on space in city centres sees buildings getting taller. From the 1860s to the 1930s engineers sought to extend the envelope for skyscrapers. In 1931 the Empire State building, at a height of 443 metres, established a standard beyond which it was difficult to go with the building technology of that time. In the last decades of the last millennium, computer analysis, computer simulation and computer-assisted design, along with improvements in concreting, welding and the quality of steel, made a new generation of high-rise buildings possible. After the events of 11 September 2001 many thought that skyscrapers might go the way of Zeppelins. Instead, the need to counter such dangers inspired new factors in design that extended the ceiling for skyscrapers. Hardened-concrete cores protect elevators and stairways from fire and blast damage. Structural sensors monitor swaying and RFID systems improve security.

The higher the building, the further the vertical distance people have to traverse to get to the ground floors where they can access road and rail transport networks. Helicopter pads on the top of a skyscraper may serve travellers wanting to get to a city airport, but the majority of people using high-rise buildings go to and from them by ground transport.

There is another problem. Many skyscrapers have developed as single purpose structures. They are hotels, or offices, or apartments or shopping complexes. Each morning the elevators of tall buildings that house apartments or hotel accommodation are packed with people trying to get to ground level where they pour onto the streets to fill the sidewalks, metros and bus routes as they hurry to choke the elevator systems of the skyscrapers where they work or shop or eat. Every evening this movement is repeated in reverse and energy consumption grows apace.

Logic would suggest that tall buildings should be either multipurpose precincts where people can live, work, shop and play in the same building or neighbouring buildings that are interlocked by bridges to allow horizontal transport between skyscrapers at multiple levels.

Moshe Safdie outlined an idea for a new type of urban complex, where towers were linked at several levels (Safdie and Kuhn, 1998). What we might think of as first generation high-rises have two planes of activity, the vertical of tall buildings and the horizontal of street level. Second generation very high buildings could have levels that give access to underground or overhead railway networks or urban freeways or walkways. If Safdie's idea were applied to a metropolis of seriously tall buildings they could be interlinked at multiple levels to provide a multi-dimensional urban transport matrix. This would make possible buildings that broke through the 1 kilometre barrier to reach 5 kilometres or more in height and house a city in a single building. Today's structural materials may not be able to stand the pressures and strains that would be involved in such giant structures, but materials made with nanotechnology would be lighter and stronger. Nanotechnology also makes it possible to think of some overall protective covering for the matrix that would provide climate control. This could become increasingly critical as global warming continues and could be a precursor to the kind of building technology that will be needed to colonize planets. Nanosurfaces would generate solar power and terraces could allow the barren sides of buildings to be developed as hanging gardens maintained by hydroponics.

The scale and structural strength of a matrix city made with nano-technology would allow the top surface of the matrix to be used as an airport. Kasarda (2001) argues the case for an aerotropolis that would refocus the life of a metropolis around an airport in the way seaports and railway stations did in industrial societies. The compacting of airport with metropolis in a structural matrix would resolve the ground transport gridlocks that exist between airports and urban areas.

Figure 13.2 The problem with first generation high-rise buildings in Singapore

There is no connection between these buildings other than at ground or below ground level.

Figure 13.3 The Twin Towers building in Kuala Lumpur

Multifunctional and linked above ground by a walk-through, this can be considered an example of second generation high rise.

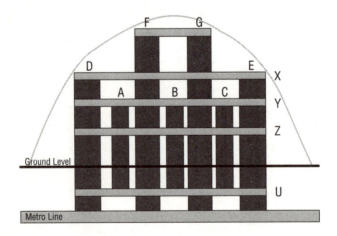

Figure 13.4 The matrix city

A, B and C are first generation high rises that have been linked by walk-throughs to third generation high rises D, F, G and E to form a matrix. The sides could have hanging hydroponic gardens and the top could be an airport.

Seacities and skycities

The US military's approach to a transport problem is to have complete control of the territory involved so that a comprehensive solution can be applied. The need for this became apparent when the United States invaded Somalia. Attention at that time turned to the idea of a mobile offshore base, a floating island that could be towed to the vicinity of a trouble spot. It would safely house stores, supplies and troops and serve as a landing strip, hospital and recreation area. From the research involved has come the idea of floating building complexes. Current designs are for residential blocks, hotels and casinos, but the basic idea could be extended to building floating cities. They could be positioned close to a coast and so be partially dependent on the infrastructures of adjacent countries. Alternatively, they could move into international waters under their own power, with their own infrastructures made with nanotechnology, able to draw on the water that supports them for their needs and able to grow and maintain themselves independently of any shipyards or any need to connect to land.

Nanotechnology will also make possible a new generation of rockets. Lighter and stronger than anything we have now, they will be capable of payloads that allow nanostructures to be built in space. It becomes possible to think of skycities in orbit around the earth or on other planets or drifting through space on great wings that catch the solar winds.

Powered pedestrians

The next great change in transport could be in the personalized transport that nanofabricated clever clothes make possible. People will be able to run, jump and swim far further and faster than is now possible and lift and carry heavy loads with ease. Elderly people, and it is likely that there will be many more of them, will be supported and assisted in their movements by their clothes so that they will have greater mobility and agility than is their lot now. At the urban, site, space and surface levels, personal movement and the transport of things by carrying them will, for the first time since humans began to walk, undergo a paradigm shift. This will not only be in the way people travel through physical reality. Clever clothes make virtual travel also possible. The growing ubiquity of small cameras is putting the infrastructure for this in place.

Wherever there are fixed surveillance cameras there is the possibility to make a 3D virtual replica of what lies within range of the lens and for people to visit the site as telepresences (Ahuja and Sull, 2001).

Humans cannot fly like birds because they lack wings, the strength to control wings and the skills of flying. Nanotechnology could be used to create foldback, extendable wings that would be lighter than the clothes we wear. They would be strong, powered partly by people, partly from an auxiliary power pack on the back and with built-in automatic flight controls. Flight in the manner of birds could be possible. People could fly through the interstices of matrix cities from one skyscraper to another as casually as they now walk between buildings.

The personal pod

For many, the best moment in a long journey is when they sink into the back seat of a taxi, tell the driver where they want to go and relax while they leave all the traffic hassles to an expert. This is what self-driving cars promise, but the idea can be extended to that of a personal multimodal pod in which passengers can stay in comfort throughout a journey leaving all the hassle of switching between different transport modes and network levels to the pod.

In the role of cars, pods could be marshalled into electronically linked platoons. Unlike trains and trams that are restricted to running on defined rail routes, car pods and platoons would join, detach, rejoin and move along any urban street that had a roadbed with the necessary communications system. There would be no more crashes at intersections because fallible human drivers failed to read or interpret signs and signals properly. Running red lights would be impossible. Personal pods would give way and avoid conflicts with pedestrians and cyclists. Segregation of traffic would be achieved not by assigning exclusive lanes to particular classes of vehicles but by protecting the space a pod is about to occupy. Priority could be assigned to emergency vehicles or to any class of vehicle for which a priority clearway is required. After delivering its passenger, a pod would park itself in an intelligent parking space that beamed availability to potential customers.

Personal pods need not be confined to roads. Nanofabricated and therefore light and flexible in shape, they could drive their passengers through an airport and onto an aeroplane where it would park itself and its passengers until it was time to drive off and complete the journey. This is a pod as an intermodal, taxi-type device. However, pods could be more in the nature of a houseboat or camper van, an intermodal mobile home that could travel anywhere in any mode, equally at home on land, water or in the air.

Multimodal freight pods could move through supply chains in pathways that have been optimized by intelligent communications systems. Terminals for the interchange of freight will become highly automated and devoid of humans. Goods will be tracked from many points of origin, through a multiplicity of value-adding processing and assembly plants, electronically documented and sealed for security inside containers that can be used for distribution by all modes of transport to any part of the globe. Trace-back would be enabled through communications systems that are integrated and global. Crossover conflicts will be avoided with piped transport systems that create a unique routing through different network levels. These could be virtual pipes based on GPS allocated by a central control system or vacuum propulsion systems that moved pods from one site to another – an old technology seen in department stores – but extended and made more efficient.

Which scenario?

We need a new transport paradigm, but these futures scenarios seem so like science fiction that anyone involved in the design and planning of transport would surely hesitate to put their professional reputation at risk by advocating any of them. How do we evaluate them?

Research in transport is today dominated by an empirical approach that in practice focuses on individual factors in transport systems. Such research is within the parameters of the existing paradigm and favours the *status quo*. It also fails to take into account the way transport is made possible by the different kinds of network systems (NETS), something that any serious attempt to plan a paradigm shift in transport must do. For example, if road transport were to be controlled and driven by AI expert systems instead of humans, then the signage that is visible and audible to humans would need to be changed to one of radio interaction between traffic and between traffic and the road infrastructure. The dynamics of the traffic network and the road infrastructures would then be changed as cars were able to close up and react to over-the-horizon conditions. This in turn would mean that the regulations governing the use of roads would need to be revised. This suggests that any approach to new transport design and planning issues should have a multidisciplinary team of experts in the different NETS.

Transport research tends to be directed towards national issues, but changes in transport must now take place within the context of global intermodal supply chains. NETS are found in every mode and at every fractal level of transport, so that planning teams in transport must also take these dimensions into account. This begins to sound like a case for a new bureaucracy, but the International Telecommunications Union (ITU) has shown how such multinational, multiskilled study groups in combination with advisory groups can work. This is the research approach behind the efficient running and development of the world's biggest machine (the public telephone system) and behind the paradigm shift to the internet.

The global episteme

The episteme we are in belongs to the United States. Currently it is the dominant culture and exerts a global hegemony with its control of transport and communications. It is the country of the automobile, the aeroplane and the atomic bomb. More goods and people pass through its seaports and airports and along its roads than in any other country. It is the transport centre of the world. The United States also controls the internet and much of the world's telephony and is the centre of the computer industry. Hollywood feeds the imagination of the world so that everyone carries a common set of visual concepts. The global language is the language of the United States. It is the communications centre of the world.

Will the United States still be the dominant empire of the next episteme or will the next world language be Mandarin, Hindi or Arabic? Or is the United States the last empire? If globalization is to be an orderly rather than a chaotic condition, it needs a paradigm shift in governance from the national to the global level: not the parliament of nations that we have now in the UN, but a sovereign federal global government democratically governed by the freely elected representatives of all people. How else can we survive as developments in transport and communications relentlessly draw us together and we face the consequences of global warming?

O'Hagan (1996) argued that if transport were to resolve the problems of distance, so that anyone anywhere could meet with and deliver products to anyone anywhere, we would then realize that what kept the peoples of the world apart was not physical distance, but differences in language and culture. O'Hagan coined the term 'teletranslation' to describe the emerging technology that allows people who speak different languages to communicate by telephone or the internet and be mutually intelligible. The next 20–30 years will see the development of IT to the point where anyone anywhere could attend a session of a world parliament in HyperReality, understand what is being said, and vote on it. Which of the scenarios outlined above would they then vote for?

References

Ahuja, N and Sull, S (2001) Hypervision, in *HyperReality: Paradigm for the third millennium*, eds J W Tiffin and N Terashima, pp 43–53, Routledge, London

Amon, C H *et al* (1997) Thermal management and concurrent system design of a wearable multicomputer, *IEEE Transactions on Components, Packaging, and Manufacturing Technology*, **20** (2), pp 128–137

Ancel, M L *et al* [accessed February 2003] Applying network theory to epidemics: control measures for mycoplasma pneumoniae outbreaks, *Emerg Infect Dis* [Online] http:/www.cdc.gov/ncidod/EID/vol9no2/02-0188.htm

Ashby, W R (1991) Requisite variety and its implications for the control of complex systems, in *Facets of Systems Science*, ed G J Klir, Plenum Press, New York

Barabasi, A-L (2002) *Linked: The new science of networks*, Perseus Publishing, New York

Barker, J A (1993) *Paradigms: The business of discovering the future*, Harper Business, New York

Barthes, R (1967) *Elements of Semiology*, Hill and Wang, New York

Barthes, R (1977) *Image-Music-Text*, Fontana, London

Barthes, R (1995) *Mythologies*, Hill and Wang, New York

Berger, C (2005) *Wayfinding: Designing and implementing graphic navigational systems*, Rotovision, Hove

Bertalanffy, L von (1951) General systems theory: A new approach to the unity of science, *Human Biology* 303–361

Bhagwati, J (2004) *In Defence of Globalization*, Oxford University Press, Oxford

Birdwhistell, R (1970) *Kinesics and Context*, University of Pennsylvania Press, Philadelphia, PA

Bishop, F (1996) *A Description of a Universal Assembler*, IEEE International Joint Symposia on Intelligence and Systems, Rockville, MD

Bollen, F *et al* (2004) Sea and air container track and trace technologies: analysis and case studies, Final report, July 2004 for APEC Transportation Working Party [Online] http://trackandtrace.org.nz

Bormann, E (1985) *The Force of Fantasy: Restoring the American dream*, Southern Illinois University Press, Carbondale, IL

Brand, S (1987) *The Media Lab: Inventing the future at MIT*, Penguin Books, New York

Briault, E W H and Hubbard, J H (1957) *An Introduction to Advanced Geography*, Longman, London

Brundtland, G H (2003) [accessed 24 January 2007] Day One conclusion: the response so far [Online] http://www.who.int/dg/brundtland/speeches/2003/kuala lumpur sars/en/Norton, New York

Buchanan, M (2002) Nexus: Small worlds and the groundbreaking science of networks, Norton, New York

Candemir, Y [accessed 24 January 2007] Introducing 10 [Online] http://www.wctrs.org/indexconferences.htm

Capella, Y (1988) Symposium on mass and interpersonal communication, *Human Communication Research*, **15**, pp 236–318

Capra, F (1996) *The Web of Life: A new synthesis of mind and matter*, HarperCollins, London

Carey, J W (1989) *Communication as Culture: Essays on media and society*, Unwin Hyman, Winchester, MA

Castles, S (2003) *The Age of Migration: International population movements in the modern world*, Palgrave Macmillan, Basingstoke

Charon, J M (1992) *Symbolic Interactionism*, 4th edn, Prentice-Hall, Englewood Cliffs, NJ

Chow, G *et al* (2000) Freight transportation planning and logistics, *Transportation in the New Millennium: State of the art and future directions*, US Transportation Research Board, National Research Council, Saunder School of Business Research, University of British Columbia, Vancouver, Canada

Coyle, J J, Bardi, E J and Novack, R A (2006) *Transportation*, 6th edn, South Western College, San Diego, CA

Davidsson, E (2005) [accessed 22 January 2007] UN Secretary General Kofi Annan: Public Relations Officer for the new world order [Online] http://www.globalresearch.ca/index.php?context=viewArticle&code=DAV20050315&articleId=452

Deloitte (2004) [accessed 23 January 2007] Prospering in the secure economy: a Deloitte research study [Online] http://www.deloitte.com/dtt/cda/doc/content/Prospering%20Australia%20-%20low%20res.pdf

Drexler, E K (1986) *Engines of Creation*, Anchor Books, New York

Ekman, P and Friesen, W V (1975) *Unmasking the Face: A guide to recognizing emotions from facial clues,* Prentice-Hall, Englewood Cliffs, NJ

Eschle, C (2005) *Critical Theories, IR and the 'Anti-globalization' Movement,* Routledge, London

Everett, J L (1994) Communication and sociocultural evolution in organizations and organizational populations, *Communications,* **4**, pp 93–110

Fayol, H (1949) *General and Industrial Management*, Pitman, London

Festinger, L (1957) *A Theory of Cognitive Dissonance,* Stanford University Press, Stanford, CA

Fisher, B A (1970) Decision emergence: phases in group decision making, *Speech Monographs,* **37**, pp 53–60

Forgacs, D ed (2000) *The Antonio Gramsci Reader: Selected writings 1916–1931,* New York University Press, New York

Foucault, M (1972) *The Archaeology of Knowledge*, Pantheon, New York

Foucault, M. (1973) *The Order of Things,* Vintage Books, New York

Gagne, R M (1965) *The Conditions of Learning,* Holt, Rinehart and Winston, New York

Giarratano, J and Riley, G (1998) *Expert Systems: Principles and programming,* PWS Publishing Company, Boston, MA

Gibson, W (2003) *Pattern Recognition,* Viking, London

Goffman, E (1974) *Frame Analysis: An essay on the organization of experience,* Harvard University Press, Cambridge, MA

Golledge, R G ed (1999) *Wayfinding Behaviour: Cognitive mapping and other spatial processes,* The Johns Hopkins University Press, Baltimore, MD

Hall, E (1959) *The Silent Language,* Fawcett, Greenwich, CT

Hall, J S (1999) [accessed 24 July 2003] Architectural considerations for self-replicating manufacturing systems, *Nanotechnology* [Online] http://www.foresight.org/Conferences/MNT6/Papers/Hall/index.html

Hawkins, R P, Wiemann, J M and Pingree, S eds (1988) *Advancing Communications Science,* Sage, Newbury Park, CA

Heath, R L and Jennings, B (2000) *Human Communications Theory and Research: Concepts, context and challenges,* Lawrence Erlbaum, Mahwah, NJ

Heinich, R (1970) *Technology and the Management of Instruction: Monograph 4*, Association for Education Communications and Technology, Washington, DC

Hibbs, J (2000) *An Introduction to Transport Studies*, Kogan Page, London

Hickman, C A and Kuhn, M (1956) *Individuals, Groups and Economic Behaviour*, Holt, Rinehart and Winston, New York

Humphreys, I and Ison, S (2005) Changing airport employee travel behaviour: the role of airport surface access strategies, *Transport Policy*, **12** (1), pp 1–9

IPCC (2001a) *The Regional Impacts of Climatic Change*, Cambridge University Press, New York

IPCC (2001b) *Special Report on Emissions Scenarios*, Cambridge University Press, New York

Jackson, P (1999) *Introduction to Expert Systems*, Addison Wesley, Harlow

Juels, A, Molnar, D and Wagner, D (2005) [accessed 23 January 2007] Security and privacy issues in e-passports [Online] http://eprint.iacr.org/2005/095.pdf

Kansky, K J (1963) *Structure of Transportation Networks*, The Department of Geography, University of Chicago, Chicago, IL

Kasarda, J D (2001) From airport city to aerotropolis, *Airport World I* **6** (4), pp 42–45

Katz, E, Blumler, J G and Gurevitch, M (1974) Uses of mass communication by the individual, in *Mass Communication Research: Major issues and future directions*, eds W P Davidson and F Yu, pp 11–35, Praeger, New York

Kerouac, J (1958) *On the Road*, Deutsch, London

King, D (2002) [accessed July 2006] The science of climate change: adapt, mitigate or ignore? [Online] http://www.foundation.org/UK/801/31102 2pdf#search=%22Ninth%20%Zuckerman%20lecture%22

Koen, F (2006) *Innovation, Evolution and Complexity Theory*, Edward Elgar, Cheltenham

Kolbert, E (2006) *Field Notes from a Catastrophe*, Bloomsbury Publishing, London

Kreps, G L (1990) *Organizational Communication Theory and Practice*, Longman, New York

Kuhn, T S (1962) *The Structure of Scientific Revolutions*, University of Chicago Press, Chicago, IL

Lambert, D M, Stock, J R and Ellram, L M (1998) *Fundamentals of Logistics Management*, Irwin/McGraw Hill, Boston, MA

Lasswell, H (1948) The structure and function of communication in society, in *The Communication of Ideas*, ed L Bryson, Institute for Religious and Social Studies, New York

Laudon, K C and Laudon, J P (2005) *Management Information Systems: Managing the digital firm*, Prentice-Hall, Upper Saddle River, NJ

Littlejohn, S W and Foss, K A (2004) *Theories of Human Communication*, Wordsworth, Belmont, CA

Mandelbrot, B B (1982) *The Fractal Geometry of Nature*, W H Freeman, San Francisco, CA

Mandelbrot, B B (2006) *The (Mis)behaviour of Markets: A fractal view of risk, ruin and reward*, Perseus Books Group, New York

Marx, K (1859, 1977) *A Contribution to the Critique of Political Economy*, Progress Publishers, Moscow

McColgan, M D (2001) *The Cold Chain*, Lincoln University, Lincoln, NZ

McCulloch, W S (1988) *Embodiments of Mind*, MIT Press, Cambridge, MA

McLuhan, M (1964) *Understanding Media*, McGraw Hill, New York

McNurlin, B C and Sprague, R H (2005) *Information Systems Management in Practice*, Prentice Hall, Harlow

Merkle, R C (1997) [accessed 24 July 2003] Convergent assembly, *Nanotechnology*, [Online] http://www.zyvex.com/nanotech/convergent.html

Milgram, S (1967) Small world problems, *Psychology Today*, pp 60–67

Mitchell, S A and Black, M J (1996) *Freud and Beyond: A history of modern psychoanalytic thought*, HarperCollins, London

Moreno, J (1934) *Who shall Survive?*, Beacon Press, New York

Morlock, E (1967) *An Analysis of Transport Technology and Network Structure*, The Transportation Centre, Northwestern University, Evanston, IL

Narayanan, S (1997) [accessed 23 January 2007] Talking the talk is like walking the walk: a computational model of verbal aspect, CogSci97 [Online] http://www.icsi.berkeley.edu/NTL/papers/sub.pdf

Nickel, S and Puerto, J (2005) *Location Theory: A unified approach*, Springer, New York

O'Hagan, M (1996) *The Coming Industry of Teletranslation*, Multilingual Matters, Clevedon

O'Hagan, M and Ashworth, D (2002) *Translation-mediated Communication in a Digital World: Facing the challenges of globalization and localization*, Multilingual Matters, Clevedon

Ogden, C K *et al* (1923) *The Meaning of Meaning: A study of the influence of language upon thought and of the science of symbolism*, Kegan Paul Trench Trubner, London

Pearce, F (2006) State of denial, *New Scientist* **4** (November), pp 18–21

Perez, C. (1983) Structural change and the assimilation of new technologies in the economic and social systems, *Futures*, **15**, pp 357–375

Perl, R (2006) *Trends in Terrorism: 2006*, US Foreign Affairs, Defense and Trade Division, Congressional Research Service (US)

Phoenix, C (2003) Design of a primitive nanofactory, *Journal of Evolution and Technology*, **13** (2)

Poirier, C (2006) *RFID Strategic Implementation and ROI: A practical roadmap to success*, J Ross Publishing, New York

Popper, K (1959) *The Logic of Scientific Discovery*, Hutchinson, London

Posner, R A (2004) *Catastrophe: Risk and Response*, Oxford University Press, New York

Rahman, S and Findlay, C (2003) The impact of terrorism on cycle times in international supply chains, *Journal of International Logistics and Trade*, **1** (1), pp 44–53

Richardson, M. (2004) *A Time Bomb for Global Trade*, Institute of Southeast Asian Studies, Singapore

Rimmer, P J (2003) The spatial impact of innovations in international sea and air transport since 1960, in *Southeast Asia Transformed: A geography of change*, ed Chia Lin Sen Institute of Southeast Asian Studies, Singapore

Robinson, R (2002) Ports as elements in value-driven chain systems: the new paradigm, *Maritime Policy & Management*, **29** (3), pp 241–255

Rogers, E M and Kincaid, D L (1981) *Communications Networks: Towards a new paradigm of research*, Free Press, New York

Rothengatter, T. (1997) Psychological aspects of road user behaviour, *Applied Psychology: An International Review*, **46** (3), pp 223–234

Rothengatter, T and Huguenin, D eds (2004) *Traffic and Transport Psychology: Theories and applications*, ICTTP 2004, Elsevier, Amsterdam

Royal Society, The [accessed 20 January 2007] Nanoscience and nanotechnology: opportunities and uncertainties [Online] http://www.nanotec.org.uk/finalReport.htm

Russell, P (2001) *Henry the Navigator*, Yale University Press, Cambridge, MA

Safdie, M and Kuhn, W (1998), *The City after the Automobile: An architect's vision*, A New Republic Book/Basic Books, New York

Schermerhorn, J R, Hunt, J G and Osborn, R (2004) *Organizational Behaviour*, Wiley, Chichester

Shannon, C E (1948) A mathematical theory of communication, *The Bell System Technical Journal*, **27**, pp 379–423, 623–656

Shannon, C E and Weaver W (1949) *The Mathematical Theory of Communication*, University of Illinois Press, Urbana, IL

Sjosted, L (2002) Managing sustainable mobility: A conceptual framework, in *Information Systems in Logistics and Transportation*, ed B Tilanus, pp 9–10, Pergamon Press, Oxford

SmartCommunitiesNetwork (2005) [accessed 22 January 2007] Topics in sustainability [Online] http://www.smartcommunities.ncat.org

Spoor, M ed (2004) *Globalization, Poverty and Conflict: A critical development reader*, Kluwer Academic Publishers Group, Dordrecht and Boston, MA

State Department (US) (2006) Country reports on terrorism [Online] http://www.globalsecurity.org/security/library/report/2004/pgt_2003/index.html

Stock, J R and Lambert, D M (2001) *Strategic Logistics Management*, McGraw-Hill/Irwin, Boston, MA

Taylor, F W (1911) *Scientific Management*, Harper & Row, New York

Taylor, J R *et al* (1996) The communicational basis of organization: between the conversation and the text, *Communication Theory*, **6**, pp 1–39

Taylor, M C (2003) *The Moment of Complexity: Emerging network culture*, The University of Chicago Press, Chicago, IL

TCU [accessed 14 January 2007] History of Transportation-Communications International Union [Online] http://members.aol.com/tcucarmen/tcuhist.htm

Terashima, N. (1993) *Telesensation: A new concept for future telecommunications*,TAO First International Conference on 3D Image and Communication Technologies

Terashima, N (2001) The definition of HyperReality, in *HyperReality: Paradigm for the third millennium*, Routledge, London

Tiffin, J and Rajasingham, L (1995) *In Search of the Virtual Class: Education in an information society*, eds J W Tiffin and N Terashima, pp 4–24, Routledge, London

Tiffin, J and Rajasingham, L (2003) *The Global Virtual University*, RoutledgeFalmer, London

Tiffin, J and Terashima, N eds (2001) *HyperReality: Paradigm for the third millennium*, Routledge, London

Turban, E *et al* (2002) *Electronic Commerce 2002: A managerial perspective*, Pearson Education, Upper Saddle River, NJ

Watson, J and Hill, A (1996) *A Dictionary of Communications and Media Studies*, Edward Arnold, London

Watzlawick, P, Beavin, J and Jackson, D D (1967) *Pragmatics of Human Communication: A study of interactional patterns, pathologies and paradoxes*, Norton, New York

Weber, M (1948) *The Theory of Social and Economic Organization*, Oxford University Press, New York

Weick, C (1979) *The Social Psychology of Organizing*, Addison-Wesley, Reading, MA

Wenzel, P (2005) *Signage Planning Manual*, Lulu Press, Napa, CA

White, L (1966) *Mediaeval Technology and Social Change*, Oxford University Press, New York

White, M (2005) [accessed 24 January 2007] Source list and detailed death tolls for the twentieth century hemoclysm [Online] http://users.erols.com/mwhite28/warstat1.htm

Wiederkehr, P (2002) [accessed 8 April 2007] Environmentally sustainable transport in the CEI countries: concept, strategy and long-term benefits, CEI Summit Economic Forum, Skopje [Online] http://www.ceinet.org/download/sef_2002/22_Wiederkehr.pdf

Wiener, N (1948) *Cybernetics or Control and Communication in the Animal and the Machine*, MIT Press, Cambridge, MA

Wohl, R (2005) *The Spectacle of Flight: Aviation and the western imagination 1920–1950*, Yale University Press, Cambridge, MA

Wolf, M (2004) *Why Globalization Works*, Yale University Press, Cambridge, MA

WRI [accessed 20 January 2007] World Resources 1998–99: Environmental change and human health [Online] http://pubs.wri.org/pubs_description.cfm?PubID=2889

Index

11 September 2001   1, 12, 96, 105, 164,
      175, 178, 188, 190, 200
1960s, mutually reinforcing
      paradigm shifts in transport and
      communications   193
200 nautical mile economic exclusion
      zone   121
2D images   101
3D virtual reality   101
3D virtual replica   204

abstract networks   46
      linked by logic and antecedence   46
abstract systems of a communicative
      nature   38
access to airports   186
accidents from failure to observe
      signs   62
accounting   169
acoustic range of human voice   145
action research favours existing
      paradigm   205
activity spaces
      allow face-to-face communications
         149
      common denominator are the doors or
         gates to allow entry or exit   149
      dedicated by purpose   148
activity spaces typified by rooms   148
acts of terrorism are acts of
      communication   178
Adam Smith   78, 194
adding value   22
adding value to transport   124
addresses, having an address   131, 132
administrative theory   18
advertise   46
aeroplane, moving string of rooms   153
aerotropolis   201
aesthetics of transport
         communications   71
after delivery communications   168
AI expert systems   205
AIDS   183
air and water, volatile mediums   181
air freight storage   165
air pollution, buildings as sources   154
air rage   189
air services, bilateral agreements   125
air space, international agreements   118
air traffic control   116, 167
      harmonization with few centres   118
air transport   104, 107
      domestic   112
      fractal levels   111
air travel   107
      safe form of transport   188
airborne restaurants, theatres,
         gymnasia   125, 126
airbridges   116

Airbus A380   111
aircraft
  air circulation of bacteria and
    viruses   188
  increasing size   113
aircraft carrier, floating airport   153
aircraft cockpit displays   167
aircraft food and beverage services   188
aircraft maintenance   98
aircraft noise   113
aircraft seating
  direction   188
  space allocation   188
airline alliances   125
airline as communications network   124
airline as TSP   124
airline bankruptcies   109
airport check in   186
airport concessions
  like passengers to spend more time in
    airports   187
  source of revenue for airports   187
airport ground control   167
airport location   115
airport security   112
airports
  as shopping malls   188
  downtown   122
  international   112
airships   108
airspace, sovereign jurisdiction   121
alternative energy sources   195
'amber' routes, Mediterranean to
  Baltic   130
ambiguity, ways to reduce   74
Amsterdam   110
answering questions not readily done by
  computers   138
Antarctica ice research   182
antibiotic resistant superbugs   183
anti-globalization   68
arbitrary arrest   179
archipelagic countries, ferry
  services   110
architecture   95
Argentina   179
artificial intelligence (AI)   preface, 12,
  43, 46, 48, 50, 99, 101, 103, 106, 120,
  196
  definition   103

managing roads   105
researchers   103
road traffic drive itself   142
robots   104
strong   103
traffic control   105
artificial life   106
assessment of security threats   167
Association of International Automobile
  Manufacturers (AIAM)   9
assumption movement of information
  and goods is the same   11
atmospheric energy levels   182
atmospheric impairment to radio
  communications   117
augmented reality   12
Australia   195
automania   185
automatic check in for flights   186
automatic doors   41
automatic pilot   20, 119
automatic turnstiles   41
automobile accidents   185
automobiles
  dangerous   134
  expensive   134
  inefficient   134
  major factor in global warming   134
automobiles and aircraft construction
  with nanotechnology   197
autonomous automated cars   196
autonomous unit   41
auxiliary service provider (ASP)   47, 92,
  124, 136, 137
auxiliary services   124
avalanche   181
aviation, experimentation   109
aviation services, notional lanes   114
avoidance of apprehension and
  interrogation   168

baggage   153
balance between first and second order
  communications   73
Balkans   179
ballast water   120
Baltic   109
Bangkok   110
Bangladesh, monsoon floods 2004   190
banking   124

banks   164
barcodes   144, 165, 166, 174
basic communication systems enables
      NETS to function   46
basic functions, sending (dispatch) and
      receiving   41
basic relationship transport and
      communications   4
basis for control   50
battleships   180
beacons   114
before transport planning   163
behavioural norms created by
      traffic   140
behavioural sciences   13
Bell, Alexander Graham   84
bell curve   31
belligerent non-nuclear countries seek
      nuclear weapons   180
benchmarking   169
Berlin wall, collapse   78
berthage, guaranteed   118
Bertrand Piccart   108
Bicycle, a moving seat   153
billing   169
bills of lading   165
binary digit   8
biometric information and security   186
Bluetooth environment and
      networks   98
boat people   190
boat people   110
bodies of knowledge as abstract
      networks   38
body condoms   102
body language   74
Boeing   777ER   111
booking systems, computer based   124
brand image   171
branding   174
Brazil   194
British Navy   3
broadband   Preface
broadband capability   91
broadband internet   197
bubonic plague   183
building by accretion of atoms   97
buildings interlocked by bridges at
      multiple levels   200
buildings that use IT = smart

buildings   155
bulk carriers   98, 125
bulk carriers, submersible   127
bulk transport
      decline   98
      of raw materials and finished
            goods   196
buoys   114
bureaucracy theory   17

calamity, scope for   42
cameras and sensors   105
canalized rivers   115
canals, narrow   110
Cape Horn   109
Cape of Good Hope   109
captain's table   120
car
      advertising   185
      advertising with health warning   185
      dependency   185
      driving skills   48
      parking   185
      platoons locked electronically   133
      small room that moves   153
carbon capture and storage   195
carbon credits   195
carbon dioxide   181
carbon-based fuels   92
cargo handling, reduced turn-around
      times   86
cargo-handling systems   125
cars
      allow people to be in charge of own
            destiny   139
      as back-seat drivers   141
      autonomous automated   196
      medium of communicating   139
      self driving   204
      submissive servants no longer   141
      that drive themselves   136
cattle class travel   185
causal circularity in transport leads to
      globalization   77
Cecil Rhodes, dream of Cape to Cairo
      railway   130
cellular radio   98
cellular systems   163
change of gauge, airlines – inefficient
      112

Changi airport   188
    automatic check-in   187
chaos   20
cheap energy   194
check in systems at airports   186
check pilots   119
checkout counter   144
chemicals transport   190
Cheo Lap Kok international airport   165
Chile   179
China   185, 194
China Sea   109
cholera   183
chronometers   117
circular causality   19, 164
    locks together paradigms and
        syntagms   79
circumnavigation   107
    Victoria   108
cities as railway hubs   131
city centre to city centre, 500km high
        speed rail beats air transport   131
city in a single building   200
city of Ames   51
classification transport   4
clear communication   47
clear transport communications   51
clever clothes   98, 99, 100, 102, 103, 136,
        156, 196, 197
    act robotically   100
    amplify and support movement   99
    concept   102
    nanofabricated   203
clever cottages   197, 198
climate change   2, 93, 96, 181, 194
    greater threat than terrorism   177
climate control in buildings   201
clothes
    clever enhancements   99
    utilitarian to protect and
        communicate   99
clustering of paradigm shifts   1960s   91
CO2   194
CO2 emissions from transport   185
coaction field   101, 102, 104
coal and oil   194
coastal shipping   88
coastal waterways   109
code, a set of signs   54
codes and signs   54

cognitive dissonance   70
    seeking consistency   71
Cold War   180
    end of   78
commercial pilot's license (CPL)   119
common documents   165
common understanding with first order
        communications   54
communicating by dress, haircut,
        gesture, body language, scent, the
        way people walk   146
communication
    first and second order   47
    people to people   144
    two way flow in supply chains   159
communication activities depend on
        location of economic activity   33
communication methods in supply
        chains, evolving continuously   163
communication networks   46
communication nodes
    aviation   114
    shipping   114
communication paradigms impacted by
        paradigm shifts in transport and
        vice versa   89
communications
    an intermittent activity for
        humans   15
    depends on language   123
    effectiveness of   22
communications
    face-to-face   74
    highway code   140
    intra-traffic   140
    key to efficient functioning of a
        system   157
    line of sight   117
communications by car within, between,
        and with road infrastructure   139
communications compliance, use of
        computerized communications   74
communications for transport
        logistics   157
communications networks   122, 137
communications studies   14
    lacks core body of theory   82
communications systems
    in roadbeds   204
    once essentially transport systems   89

communications technologies, next
    generation   Preface
communications without ambiguity   52
communicative behaviour of vehicles in
    close proximity   140
communist ideologies   18
community growth and continuance,
    depends upon transport and
    communications infrastructures   11
commuter trains   131
competing standards   166
competing value-adding supply
    chains   160
complexity theory   14
compliance, export/import security   164
compliance with regulations   46
compressed air powered cars   135
compulsory education   90
computer assisted design (CAD)   199
computer based booking systems   124
computer-mediated
    telecommunications   197
computers
    automate the flow of quantitative first
        order information   144
    built-in to check performance and
        behaviour   63
    do not read road signs – yet   64
    internet-linked   99
    more powerful and interlinked   105
    no human behaviour problems   63
    not absent minded, drowsy or
        drunk   120
    shrinking in size becoming
        ubiquitous   98, 196
concept schema   4
conceptual framework   1
concrete networks of infrastructure and
    traffic   38
confidence in supply chains   164
conformation to dominant ideology   67
Congo   179
conjunction of paradigm shifts   84
conjunction of RFID, cameras, sensors,
    telecommunications, computer
    hubs   105
conjunction of transport and
    communications   2
connotation   65
container ships   98

container temperature control   171
containers
    introduction   86
    made intermodality transport
        possible   86
continental rivers and canals   109
control of mass media   67
controlling and improving networks   50
conurbations concentrate
    population   185
convergent assembly   95, 98
convergent assembly, systematic and
    hierarchical architecture   95
conveyor belts   144
coping with disease, natural disasters
    and violent people   184
cost of flying   112
    relative to wages   113
costs
    lowering costs of using
        telecommunications   199
    rising cost of using road infrastructure
        and vehicles   199
costs of accidents   136
costs of motoring, full costs ignored in
    most countries   135
cottage industry   98, 198
coupling GPS and RFID   62
critical information   116
critical theory   67, 68
crossover conflict resolution   205
cultural communications issues   114
curfews   114
customs   112, 124, 186
cybernetic control   114
cybernetic flows of information indicate
    state of network   45
cybernetics   14
    first order   20
cybernetics theory   19

danger from disease   183
danger from freak conditions, lack of
    information   182
danger in transport, ignorance   183
Darfur   179
Das Kapital   78
data errors, reduction in   175
data warehouse   169
de facto traffic behaviour   141

de jure traffic behaviour   141
decision support system   42
decisions automated   41
decisions made by people   41
decline of bulk carriage by land and
        sea   196
declining technology   89
deep linking in databases   168
definition
    globalization   9
    information technology   7
    telecommunications   2
    transport   4
demand for transport, growth
        assumption   96
denotation   65
departure, place of   15
design of transport infrastructures and
        vehicles   71
desks and chairs   149
desktop factory   98
despatching and docking   22
destination, place of   15
deterioration   22
deterioration in transit   161
developed countries   92
developing countries   68, 92
diamonoid fibres   99
differences in language, culture, political
        systems impact supply chain
        management   159
dirigibles   96
disabling cars from intoxicated or
        fatigued drivers   141,142
disappearance of dissidents   179
distal stimuli   60
distributed virtual reality   120
distribution, reduction in volume   196
DNA matching for tracing   174
dock and link   45
doctors, virtual rounds   103
documentation
    compliance   175
    electronic   116
    requirements   164
    ship's manifests   118
    streamlining   175
domain knowledge (DK)   101
domestic aviation and air services   88,
    112

dominant ideologies of society   67
domino effect   84
domino effect in transport and
        communications   92
doomsday scenario   194
doors that recognise people with right
        protocol   155
double interact   140,141
double interact as basis for organizing
        behaviour   140
downloads   197
Dr Frankenstein   103
dramas of history   12
dress-codes   165
driver factors – health, emotional
        maturity, reaction speed,
        eyesight   135
driverless cars   142
drivers
    moods and attitudes   140
    with low skills kill more people than
        terrorists   135
    as antisocial way of life, end to   142
    as communicating   139
    liked   142
driving, without physical presence   102
drogues   103
dyadic relationship   24

Earth, a finite system   193
Earth's rotation, rocket travel   126
earthquake   181
ease of obtaining right to drive a
        car   135
Ebola virus   183
e-books   195
e-commerce   164, 195
economic factors, distance   34
economic interdependency, countries less
        self-sufficient   180
education, process of learning paradigms
        that comprise epistemes   81
El Salvador   179
e-learning   195
electric vehicles   135
electromagnetic pulse (EMP)   180, 181
electromagnetic spectrum, access to   100
electronic billing   142
electronic data interchange (EDI)   163,
        185

electronic documentation   116
electronic gantries   167
electronic payment   124
elevated railways   131
elevators   86, 199, 200
e-mail   165
embedded AI   197
embedded telemetric devices   196
emergency services   136
emergency workers   100
Empire State Building   199
Empires, epistemes in expansion   85
Energy, need for   15
energy for transport   5, 96
    shifts from wind and animals to coal,
        to oil   92
energy involved in transport   134
English, language of civil aviation   123
environmentally sustainable transport
    (EST)   129
e-passports   186
episodic nature of transport
    communications   49
episteme   78, 84
Episteme
    a paradigm that encompasses all other
        paradigms   78
    current one not benign   178
    gives rise to common way of
        thinking   84
    in growing democracy   134
    refers to totality of intermeshed
        paradigms in society   84
episteme of globalization   92
epistemic change   175
epistemic shifts from revolution,
    invasion, conquest, colonization
    84
epistemic thinking   96
erasure, difficulties   74
escalator, infrastructure that moves   45
escalators   199
essential services   124
ethnic cleansing   179
European Union road and rail
    systems   130
Exclusive Economic Zone (EEZ)   121
existence and survival
    cannot without transport   15
    cannot without communications   15

experience, accumulation of   104
expert system   104, 120
expert systems that outperform
    humans   104
explosives transport   190
exponential expansion of transport
    activity   194
express trains   132

fabrication with micro-machines   97
face-to-face communications   74
face-to-face links   125
facial expressions   146
failure to observe signs   62
fantastic transport communications   65
fantasy world   76
fare collection system   113
fast traffic   132
fax-back of signatures   168
feedback   14, 19.   163
feedstock   97
feminist studies   68
ferry services   110
feudal episteme   85
fibre optic connection   197
fields of separate academic enquiry   12
fields of separate professional
    proficiency   12
first non-stop round-the world by
    balloon   108
first order communications   51, 139
    terrorism   179
first order cybernetics   163, 168, 175
    during transit   164
first order meaning   51, 123
first order of signification   52, 66
fixed schedules timetabling   118
fixed signs are 'signage'   57
fixed space   146
flag signalling   85
flight deviations   111
flight information services   167
flight simulators   119
flight timetabling   117
floating cities   203
floating islands   203
floating nodes   32, 34, 116
flood   181
flooding of flatlands   194
flow of traffic, limits to   117

fly like birds, realization of human
    fantasy 99
foldback extendable wings 204
footpaths, tracks and rights of way 131
Fordism 18
fork-lifts 144
fortress mentality 190
fossil fuels 96, 108
    reduction 98
fractal geometry 24
fractal levels 158
    aviation, third level 112
    activity spaces 148
    site 148
    site spaces and surfaces 144
fractal levels of networks 50
fractal levels of skills networks 49
fractal nature of nanotechnology 98
fractal shifts 24
fractal theory 14
fractals 24
free flow people, finance, information
    and services 9
free market philosophies in
    aviation 109
free trade 9
    maximization 194
freight hub airports 165
freight terminal automation 205
freight transport, shift from moving large
        quantities long distances to moving
        small quantities short distances 197
fuel efficiency 127
functional integration 159
fundamental laws of nature 24
furniture 149
future
    the road moves and traffic is
        stationary ? 45
    use computers instead of humans to
        drive vehicles 63
future for land transport 142
future scenarios 194
future transport technology, already
    invented 96

Galileo 84
genetically engineered virus 177
Germany, 1930s *autobahns* 132
global community 78, 92

global episteme 206
global hegemony 206
global intermodal supply chains 206
global mobile satellite communications
    (GMSC) 117, 118
global population 194
global positioning system (GPS) 62, 91,
    136, 167, 205
    GPS satellites 114
    precise navigation in 3D space 117
global supply chain management,
        integrates all people involved 159
global supply chains 157
    involves all fractal levels in
        transport 163
global tracking, unfulfilled quest 168
global trade, ways to increase 77
global transport
    cessation 184
    tools of oppression 68
global transport service provider 158
global village 156
global warming 5, 92, 111, 175, 177, 181,
    182, 185, 190, 194, 206
globalization 2
    began with water-borne
        transport 108
    no physical expression 9
globalization as a phenomenon 9
globalization of communications 178
globalization of transport 178
goals 19
goal-seeking systems 11, 20
good theory, testable by argument 17
grade separation 133
grammar of Highway Code 80
gratification theory 69
great circle routes 111
Great Depression 78
Great Lakes 115
greenhouse gasses 181
gridlock 62, 178

handshake, automatic 152
harbours, small 110
hard copy messages 165
harmonization 71
health-care worker vulnerability 184
hegemony 67
height of buildings 199

helicopter pads on top of buildings   200
helicopters   122
helipads   122
helium   108
Henri Giffard   108
Heraclitus   35
high speed rail track   133
high speed roads, expressways,
     freeways, motorways, *autobahns,*
     autostrada   131
Highway Code
   complexity   62
   first order of observance   141
Highway Codes   136
Hindenburg   96, 108
historical flow trace back   159
historical information,
     accumulation   167
history of communications
   extending access to and storage life of
        information   7
   extending the distance   7
   increasing the speed   7
Hollywood   206
Hong Kong   114
human communication,
     uncertainties   159
human communications   114
   distinct from IT   5
   hierarchical   6
   interaction with environment   6
   loss of   138
   processing and storing, thinking and
        remembering   5
   starts and finishes in human heads   5
human error   103
human feelings   138
human genome   33
human intelligence (HI)   101, 104, 120
human perceptual ability   38
humans as buttons that computers
     press   106
Hurricane Katrina   181, 191
hybrid powered vehicles   135
hydrogen energy   195
hydrogen powered cars   135
hydroponics   201
HyperReality (HR)   Preface, 12, 95, 100,
     101, 102, 104, 120, 196, 197, 207
   like telephoning a 3D version of

     yourself along with the place
     you are in and the things you are
     handling   102
HyperWorld (HW)   101

iconic signs   54
   transcend language
   are universal   55
imagination   47, 112, 124, 186
immigration port   151
imprisonment without due law   179
improved transport facilitates
     pandemics   183
in-car computers, linked to
     communications   141
inconsistent communications, hard for
     computers   142
indexical signs   54
India   185, 194
indicators of chaos   193
individuals, messages and meanings   52
induced stress conditions   119
industrial management   17
industrial revolution   84
   resulted from paradigm
        shifts in transport and
        communications   193
industrial society, functions   98
industrialized countries   92
informal space   146
information
   cross referencing   125
   distribution   120
information and redundancy   56
information breaks   169
information database   167
information repetition   56
   to overcome distortion   57
information revolution   84, 193
information society   196
information technology   162
information theory   14, 20
   origins   21
infra means 'underneath'   44
infrastructure
   artificial system designed, built
        and maintained in response to
        needs   44
   collapse   178
infrastructure channels traffic   43

infrastructure network, consists of fixed
facilities   45
infrastructure networks
air   113
water   113
infrastructure systems, construction
of   102
infrastructure that controls traffic   133
infrastructures are dynamic over
time   45
inland waterways   88, 109
innovative technology   89
inputs   19
institutionalization of network paths
over time   43
instrument rating   119
insurance   124
integrating intelligence, two
dimensions – how to do and actual
application   48
integrative intelligence in NETS   43
intellectual capability and signs   52
intelligence, aspects of   104
intelligence in traffic, could become
function of AI and IT   46
intelligence within NETS, people   43
intelligent parking   204
intelligent transport system   106
intentions information   116
interact   140, 141
interconnection typologies, bus, ring, star
patterns   24
interlining   121
interlocking shareholding   125
intermediaries, bypassing   159
intermodal interchange of container   171
international air services   112
International Air Transport Association
(IATA)   121
ability to collect data and monitor
global travel   121
international airports   112
international aviation
bilateral agreements   109
reciprocity   109
International Civil Aviation Organization
(ICAO)   9, 121
international hub airports   186
International Maritime Organization
(IMO)   9, 121

international space law   121
internet   12, 20, 24, 33, 91, 99, 124, 125,
152, 165, 195, 207
hunt for customers   125
hunt for services   125
linking with radio to become location
free   91
subsuming other communications
media   91
use during pandemics   184
internet access, fast   197
internet address   137
internet EDI   163
inter-operability   171
interurban transport, staged horse-
drawn   87
in-transport communications   164
intrinsic motivations   40
inventory of   8,000 signs   51
invisibility   100
invisible made visible electronically   167
IT   34, 40, 43, 46, 63, 66, 76, 98, 104, 114,
116, 136, 143, 151, 159, 180, 207
appliances   196
broadband   196
could rival or surpass human
communications in the future at
first order of meaning   66
creation of   20
dependency increase vulnerability to
nuclear war   180
growing involvement in transport   11
organizing flow of passengers   187
tracking   187
ubiquitous   196
IT and second order
communications   75
IT at site level   144
IT communications coveys message
globally very quickly   171
IT enabling environmental
friendliness   199
IT protocol   151
IT solutions for international
collaboration for control of spread of
disease   184
Italy 1920s, specific networks for   high
speed road traffic – *autostradas*   132
International Telecommunications Union
(ITU)   9, 206

Japan 196
Jaron Lanier 101
jet engines 88, 126
jet-lag 161
John Logie Baird 95
John Tyndall 181
joint use 45
Joseph Weizenbaum 103
journeys, begin and end on land 129
Juan Sebastian de Elcano 108
just-in-time 41, 114, 175

key performance indicators 169
kicking the oil habit 195
kidnappers, drug peddlers, paedophiles,
    bullies and terrorists 178
kinesics 146, 165
Kittyhawk 108
Kyoto Protocol 195

labour
    casualization 114
    flexible deployment 114
    replacement of unskilled labour by
        automation and multi-skilled
        labour 159
    unskilled and skilled 159
labouring practices, facilitation 99
land empires, made possible by road
        networks China, India, Persia,
        Incas 130
land transport
    by rail 129
    by road 129
land transport infrastructure 131
    rail and road networks 130
land transport infrastructure investment
    in links 131
land-bridges 115
landing lights 114
language multiplicity 123
languages and customs 92
laptop computers 34
Lassa fever 183
Lasswell's dictum 22
last empire 206
learning hierarchy 48
learning progression 119
legal access to property 131
legal requirements 164

letters of credit 164
level playing field 112
levers, pulleys, shovels as extension of
        arms 156
light rail 131
lighthouses 114
line-of-sight communications 46, 117
liner, floating hotel 153
lines of least resistance 132
link and dock 45
linkage, all transport and
        communications networks are
        ultimately linked 50
location theory 14, 34
logical dependency 38
    links specific skills 48
logisticians 160, 161
logistics, involves planning, assessing,
        coordinating 160
long-haul travel by air 113
lost tribes 92
low carbon economy scenario 195
loyalty schemes 125
luggage, limited enhancement 148
luxury leisure transport 89

machines acquiring intelligence 11
Magellan 78, 108
mail trains 90
making things directly from atoms and
        molecules 97
managed migration 190
management, uses flow of controlling
        information 45
management decision 41
management of transport
        infrastructure 45
manifests, electronic 118
man-machine terrestrial transport
        efficiency 143
manoeuvrable, large ships 125
manufacture, rigid structures 98
manufacturing, point-of-use 98
Mao Zedong's China 179
Marburg virus 183
Marconi 84
marine organisms in ballast water 120
maritime history, discovery 109
market intelligence 171
market penetration 169

Marxism
 ideas 67
 theory 69
 tradition 68
mass media 69
 advertising 46
 attending to people's needs 69
 vehicle for imposing a world view on
  the masses 69
 role in terrorism 179
Master's tickets, 'foreign going' 119
matching demand and supply, real
  time 124
McLuhan 155
m-commerce 163
measles 183
meat trade New Zealand-Europe 171
Mediterranean 109
melting ice caps 194
menace of idiosyncrasy 53
mental cognitive maps 62
meta-idea of modern
  communications 91
Mississippi navigation 115
mobile networks 98
mobile phones and telephony 34, 98,
  99, 137, 197
mobile transceivers 62
mode separation 133
models of transport
 behavioural 13
 economic 13
 engineering 13
momentum of oil energy
  infrastructure 195
monitoring 173
monitoring flight crew for alertness 167
monitoring status of cargo and
  vessel 120
mooring systems 125
more haste less speed 42
Moshe Safdie 200
motorized transport dominant urban
  mode 87
motorized transport is subsidized 135
motorways link cities and towns 131
movement and storage 14
movement information 116
movement of things 5
movement organization 123

movement tracking 123
multi-dimensional urban transport
  matrix 200
multiple presences in coaction
  fields 102
multipurpose multimodal
  superhubs 114
multipurpose precincts 200

nano solar paint 100
nanoblocks 97
nanocomputers 99
nanoelectric motors 99
nanofactories 97, 149, 198
nanoparts 97
nanopower plant 100
nanoscience 95
nanostructures built in space 203
nanosuits 99, 102,183
nanosurfaces generate solar power 201
nanotechnology Preface,12, 95, 97, 98,
  100, 156, 196, 197, 200, 203, 204
 design of manufacturing system 97
 in tag design 167
nanotechnology manufacture, stronger,
  less volume and mass, more
  components and complexity, easy
  to use, consume less energy than
  conventional manufacturing 97
nanotransceivers 100
 woven into fabric 99
Napoleon Bonaparte 108, 157
national daily newspapers 90
national interests, commercial concerns
  replace survival concerns 181
natural networks 132
natural threats, impacts on transport and
  communications 181
negative feedback 20, 193
negotiation 150
negotiation skills 114
neo-Taylorist analysis 43
NETS Preface, 157, 159, 162, 163, 178,
  205
 a kind of Rubik's cube 37
 all have paradigmatic and syntagmatic
  dimensions that must be
  compatible with each other 80
 applications of Preface
 are paradigms and have paradigms

nested within 80
as a tool Preface
as tool for designing, planning
    managing and revising supply
    chains 37
auxiliary network 37, 39
auxiliary services
    safety role 178
communications network 37, 39
exist in their own right 39
for railways closely identified with
    each other 130
for roads are loosely linked 130
fractal levels of air transport 109
fractal levels of water transport 109
if one changes the others must
    adapt 80
infrastructure for air is primarily in the
    nodes 113
infrastructure for water is primarily in
    the nodes 113
infrastructure network 37, 39
intelligence 43
regulatory network 37, 39
skills network 37, 39
three-dimensional matrix Preface
transport network 37, 39
use of to analyse functions 159
visualization of supply chain 37
without NETS transport would
    not happen in any systematic
    way 50
NETS categories may contain several
    similar networks 39
NETS people are drivers, pilots,
    pedestrians 43
network links
    air 116
    water 116
network purpose, to allow flows 151
network theory 14, 23
    links with cybernetics theory 24
    links with information theory 24
    links with organizational theory 24
    links with systems theory 24
    unifying aspect 24
networks
    are sets of interlinked nodes 24
    multi-disciplinary 23
networks enabling transport are

paradigmatic 92
networks of knowledge 38
new episteme 105, 106
new generation high-rise buildings 200
New Orleans 191
new places 126
new routes 126
new technologies 105
New York 114
Newton 84
next world language 206
Nicholas Negroponte 78
Nobuyoshi Terashima 100
nodal congestion near ports 117
nodes 31
    fixed facility 34
    nations as 112
    riparian trading countries 109
nodes as
    achieving the purpose of a transport or
        communications system 31
    acquiring the energy needed to cross
        the space between nodes 31
    floating nodes 32, 34, 116
    function of transport 32
    gate or portal 31
    source and destination 31
    storing or resting what is transported
        or communicated 31
    switching direction of traffic 31
no-fly zones 122
no-go areas for cars 134
noise 22, 54, 71, 114, 175
    aircraft 113
non-stop flights half globe 111
notional lanes, aviation 114
notional lanes, shipping 114
nuclear capable countries, behave
    belligerently 189
nuclear engineering 115
nuclear power energy 195
nuclear war 177

objective of many transport systems
    is to provide the means of
    communication 11
objectives 19
ocean-going vessels, increasing size 113
oceans as networks 109
Office of the United Nations High

Commissioner for Refugees (UNHCR) 190
oil energy infrastructure, enduring 195
oil spills 190
on street unexpected behaviour 142
one cannot not communicate 146
one cannot walk and not communicate 146
operational integration 158
operations around the clock 114
optical readers, low light problems 166
organization, happens through communications 16
organization theory 16, 18
organizational communications 71, 73, 114
organizational structure, flattened 17
organizations exist independently of people 16
Orwellian society 105
outputs 19
over-flying 112
overlap transport and communications 3
over-the horizon 206

Pacific Highway linking the Americas 130
packaging and labelling 174
packing people horizontally 125
pallets 144
Panama Canal 109
Pandemic, defined 183
pandemics 183
pandemics Preface, 175
paper transactions 164
paradigm 77, 79
paradigm
  is knowledge in the abstract organized as a system 79
  like a game 79
  must have syntagmatic dimension 82
  widely applied in many fields 79
paradigm and syntagm, relationship 79
paradigm shift, Suez and Panama Canals triggered shift from sail to steam 109
paradigm shifts 194
  aviation 88

brought about by discovery of paradox that cannot be resolved 84
  global level 88
  national level – interurban 87
  ships, oil driven 88
  site level 86
  urban level 87
paradigms
  change as they are used 79
  competing 85
  cybernetically adjust 79
  each transport mode needs to integrate its paradigm and move to overarching global transport paradigm 83
paradigms and the NETS 92
paradox precedes a paradigm shift 93
parallel processing network 99
passage and paths 148
passenger lists 188
passports, RFID tagged 186
path of minimum resistance 43
peak oil 92, 194
pedestrian transport 143
  where developments in transport most needed 144
pedestrians, nodes in pavement traffic 46
penny black stamp 90
people, could fly 204
people as floating nodes in a transport environment 137
people driven by automated systems 138
people transport
  source of income 197
  source of pollution 197
perception 59
perception of signs at different speeds 61
perfect communications, impossibility 53
perfect storm 190, 191
perishables 161
Persian Royal Road, Persian Gulf to Mediterranean and Aegean 130
personal choice, clever clothes 99
personal multimodal pod 204
personal pods 204
personal territory 146

personalized transport
    developments   203
person-to-person cellular
    telecommunications   197
pests transport   190
physical reality (PR)   101
Pilâtre and d'Arlandes   108
pilotless planes   196
pilots   102
pirates and piracy   125, 190
planning for motorways takes time   133
platoons linked electronically   204
poaching   122
pod
    shaped to hold its contents   152
    some form of container or
        housing   152
pods
    any mode anywhere   205
    intermodal taxis   205
    marshalled electronically   204
    nested like Russian Dolls   152
    standardized   152
pods and packaging   152
point-of sale data collection   165
point-of-use manufacturing   98
points in space, unique value at any
    given time   35
points of risk   163
policeman, node in an ASP   47
policing costs   136
policing of transport systems, reinforced
    because of terrorism   179
pollution reduction   135
popularity of private car   139
population densities   199
port control authority   116
port of entry, needs protocol   152
portable teletranslation   123
ports
    migration downstream to deeper
        water   113
    shallow draft   110
ports as doors   151
ports on computers   151
ports or portals where a node on one
    network gives access to another
    network   151
ports, portals and protocols   151
positional information   116

positive feedback   193
postage stamps   90
postal service   90
post-modernism communications   52
power, shift to oil   87
power grids, declining need for   98
powered pedestrians   203
practical experience, aviation   119
pre-clearance process   164
prerogative of humans   7
Prince Henry the Navigator   16
private pilot's license (PPL)   119
pro rate in aviation   113
problems with oil   92
processing   14
product lifecycles   175
profit vs ecological hazards, cultural
        issues, aspirations of developing
        countries   68
proof of delivery   168
property access enabled by local
    roads   131
propulsion, more effective   125
prosthetic devices   156
protectionism, rise of   78
protective covering of buildings   201
protocols, code of conduct to be followed
    in a network   151
provision of fuel   92
proxemics   146, 165
proximal stimuli   60
proximity, induces link
    intensification   45
psychology of perception   68
public lust for cars   134
public roads cater for mixed traffic types,
        possible chaos in competition for
        space   133
public roads link buildings   131
public transport crisis   13
purposeful communications, what is
        moved can be processed   22
purposive systems, transport and
    communications   19

quality control   43, 173
quality of service, decline at major
        hubs   186
quarantine   184
queuing for service   186

radar, standard navigational equipment
    from 1950 117
radar beacons 114
radio 91
radio beacons 91
radio control station 114
radioactive waste transport 190
rail, compete for passengers 130
railways
    created a reading public 90
    dominant from 1860s 87
    easy to regulate 136
    elevated 131
    fixed schedules 140
    for trains 130
    have own regulatory system 130
    underground 131
ramblers 131
random networks 32
read the road 80
realization of fantasies 76
records 105
reduction in number of controlling
    authorities 159
referent 54
refrigerated containers (reefers)
    stowage 173
refugee camps, breeding ground for
    terrorism 190
refugees 190
regulatory networks 46, 136
relative effectiveness of IT and human
    communications 67
reluctance to become dependent on
    public transport 134
remote sensing body temperature for
    health screening 188
removal of uncertainty of arrival 163
representations of abstract networks 38
reputation protection 174
reshipping 110
resolution of ground transport
    gridlocks 201
restricted access to high speed
    roads 133
reversals of transport patterns of
    industrial society 197
reverse logistics 165
reviewing performance in supply
    chain 174

radio frequency identification (RFID)
    62, 165, 166, 196, 200
    impact could rival that of
        containerization 166
RFID tagged passports 186
RFID tags 105, 123, 137
Rhine river 115
Rift Valley fever 183
riots in airports 71
rise of empires 11
risk, points of 163
risk assessment 164
road congestion pricing 134
road deaths as global epidemic 177
road system evolution 43
road traffic seeks organization 140
road transport
    access to a wide variety of
        services 136
    most dangerous 135
road use paradigm 142
road user charges 142
roads
    are for people and animals beside
        vehicles 130
    compete for passengers 130
    serve a variety of purposes 130
    urban level, many nodes 132
roads that move vehicles 133
robotized automated self 102
robots 12, 120
rocket ports 126
rocket propulsion 126
rocketry 197
rockets, new generation 203
rogue conditions 178
Rotterdam 115
Roundabout, magic 27
route conceptualization 62
route optimization, aviation 117
route recall 62
route visualization 62
rules and regulations 104
rules of thumb from experience and
    common sense 61
rural lifestyle 198
Russia 194
    trans-Siberian railway 130
Rwanda 179

safety helmets   136
safety monitoring   136
sail   88
sailing ships   88
Samuel Morse   90
SARS epidemic   2003   188
satellite footprint   98
satellites   182
    over-flying   121
    GPS   62
scale-free networks   32
scenarios of globalization   193
scheduled flights, increasing
        frequency   113
science as culture to which scientists
        conform   79
science as search to understand
        reality   79
science fiction   205
science is what scientists think   79
screening of luggage and cargo stowed in
        belly of aircraft   188
s-curve of innovation   91
sea level rise   111, 181, 190, 194
seaborne migration   108
seacities   203
sea-level canal, isthmus of Panama   115
seaport location   115
search engines as portals   151
seating
    aircraft seats face wrong way for safety
    shoulder harness   188
seats
    clustered in coaches   152
    pods for people   152
second generation very high
        buildings   200
second order communication   139, 165
second order communications
    as a drama   69
    as an act   69
    as fantasy   69
    as narrative   69
    terrorism   179
second order cybernetics   79, 168, 169
second order meaning   65, 123
second order of signification, roots in self
        rather than society   68
second order signification   66
    in transport communications,

advertising   67
    link to subconscious   68
second sourcing   175
security   119
security assurance   116
security requirements   164
security risk at every hand-off   171
self-administered reservations   124
self-driving cars   204
selfish competition   105
self-regulating autonomy   120
self-replicating nanoblock units   97
semaphore system   85
semi-fixed space   146
semiology   52
semiotics   52
seniority   119
sensors   120
sensory-motor controllers   143
separation of peoples, language and
        culture more than distance   207
sequestration of CO2   195
seriously tall buildings   200
sextants   117
Shannon's dyad, basic unit of all
        networks   21, 24, 180
ship stability   120
shipping lanes as links   109
short take-off and landing (STOL)   122
Sierpinski Triangle   29
sight and sound   61
signage and IT   63
signage and surveillance   123
signage heuristics   61
signage misinterpretation – impatience,
        frustration, anger, excitement,
        hyperactivity, drowsiness,
        depression   63
signage systems   45
signage systems
    advise, warn, guide, mandate   46
    evolve   64
signification, relationship between
        signifier and signified   65
signified   54
signifier   54
signposts   34
signs
    advertising   58
    big, frequent, loud, intense,

intermittent and moving to grab
	attention 61
cautionary   57
compulsory   57
confusing   58
counterintuitive   59
distracting   59
grab attention   61
iconic, symbolic, indexical   54
informative   57
link abstract networks with other
	NETS 46
meaning dependent on context   54
proliferation   58
standardized meaning   55
Silk route, China and India link with
	Europe 130
simulations, real time 3D   120
Singapore   115, 133, 154, 199
single hub networks   33
site level infrastructure, buildings, walls,
	rooms, corridors, stairs, fences 148
site level transport, still largely done by
	people 148
sites
	access is by door or gate 148
	nodes in urban road networks 148
	places of departure and arrival 148
	places where people live, work play
		and have an address 148
six communication links   33
six networks essential for transport   37
skill, steering   19
skills analysis
	basis for automating driving
		process 48
	pedestal for automation 119
skills network   48, 135
	integrates intelligence in NETS 48
	fractal levels 48
skycities   203
skycities in orbit   203
skyscrapers, mostly single purpose
	structures 200
slave trade   108
small world   37
smallpox   183
smart, defined in terms of using IT   199
smart blocks   199
smart buildings   199

smart buildings in smart blocks   154
smart cities   199
smart cities in smart countries   155
smart country   199
smart digital broadband   197
smart encryption   164
smart homes   199
smart offices   199
smart suits, duplication   100
smart villages   197
smell and taste   60
smuggling   190
social nuances   120
solar energy   195
solutions through IT   178
Somalia   203
sorcerer's apprentice   96
sound and sight   61
Soviet Union   78
Space
	the final frontier 107
	the next fractal level of transport 126
space travel   197
spaghetti junctions   132
Spanish flu   183, 184
speaking is to communication what
	walking is to transport 145
sports
	virtual play and virtual
		attendance 103
	ability enhancement 99
sports cheats, clothes   99
spy flights   121
St Lawrence river and seaway   109, 115
stability   127
staff replacement with automated
	systems 138
stairways and elevators   148
Stalin's Russia   179
standardization   125
standardized containers   171
Star Wars   180
Starship Enterprise   107
state highways   88
state of mind   78
steam engines   88, 96
steam power for printing   90
steam-powered derricks   86
steamships, aided by Suez and Panama
	Canals 109

Stevenson 84
stock levels 165
stockfeed 196
storage and safe keeping 15
storage location optimization 165
storm surges 111
structural sensors 200
subsidiary skills sequence 48
substitution of machines for humans 7
Suez Canal 109
superhubs 114
supermarket trolley, smart 165
supermarkets, navigate trolley
    baskets 144
supplemental communications 139
supply chain 158
    a sequence of transport services 158
    conduct of 162
    definition of 158
    elimination 98
    event management 171
supply chain integration, needs key
    information at first order of
    meaning 162
supply chain management
    requires communications at first
        order of meaning 160
    seeks system wide solutions 159
supply chain matrix 162
supply chain visibility 167
suppression of freedoms 179
surveillance cameras 204
sustainability of carbon-based
    fuels Preface
sustainable development of the
    planet 106
SUVs 199
symbiotic relationship of transport and
    communications Preface, 2, 104
symbolic interactionism 68, 69
symbolic signs 54
syntagm 79
syntagm of a paradigm, provides
    feedback to a paradigm 79
systematic transport 38
systematization of communications 14
systematization of transport 5, 14
systemic collapse 20
systemic growth 193
systems 19

systems analysts 169
systems theory 14, 19

tall buildings as multipurpose
    precincts 200
taste and smell 60
tax collection 142
Taylorism 18
teaching basic transport skills 183
technological extensions of pedestrian
    transport 153
technologies as extension of human
    capabilities 155
telebanking 199
telecommunications
    broadband 105
    direct control of traffic 91
    hardwire to protect from EMP 181
    improvements bring improved
        transport 91
    mobile 105
    primitive 3
    substitute and stimulus for travel 197
    ubiquitous 105
    wireless 3
telecommunications capability compared
    with human vision, hearing, touch
    and smell 146
telecommunications not involving
    transport 2, 11
teleconferencing, virtual 101
tele-education 184
telegraph 91
telegraph wires 90
teleimmersion 101
telemedicine 184
telemetric devices 196
telemetric systems, become
    ubiquitous 182
telephone 207
    primitive form of HyperReality 102
telephonic addresses displace place
    addresses 137
teleports 115
telepresence 102, 103, 137, 156, 197
teleprinter (Telex) 165
teleprofessionals 199
telesensation 101
telesensor 196
teleservices 184

teleshopping   199
teletranslation   12, 123, 207
teleworkers   199
teleworking   184
Telex   165
temperature recording   173
terminal infrastructure futures   196
terrorism   Preface, 154, 178
  defined   178
  prevention   190
terrorists   126, 190
  methods
  use transport to target transport   179
The Wealth of Nations   78
things transported can exist as a
      network   41
think global act local   10
third generation technology   99
third party services   164
ticketing   124
tickets
  electronic   121, 123
  physical   121
tidal energy   195
time and motion studies   149
timely communications   172
toll booths, automated   167
tomorrow's transport   Preface
torture   179
touch   60
tourism   108
tourists, virtual visits   103
trace-back   169, 205
track and trace   174
tracking   158
traffic congestion   185
traffic control systems   20
traffic control with AI   105
traffic creates network infrastructure
      43
traffic culture   140
traffic network, usually floating
      nodes   45
traffic networks   45, 116, 133
  comprise the vehicles that use the
      infrastructure   45
  linked traditionally by human
      intelligence and perception   46
traffic police   140
traffic psychology   68

traffic restraints, introduction   133
traffic segregation achieved
      electronically   204
traffic signage, based on heuristics   61
train, moving string of rooms   153
trains, smooth, seated, windless
      journey   90
trams   131
transatlantic cable   91
transceiver   166
  video, audio, text   99
transhipment   174
transmitting and receiving   22
transport
  a daily business   130
  an intermittent activity for
      humans   15
  as complex interaction   50
  deliberately initiated and systematized
      by humans within their world
      view at a particular period in
      time   16
  looking for solutions in new
      technology   96
  much to do with bringing people
      together in order to talk   146
  needs someone or something to do the
      transporting   40
  romanticized   69
  to happen needs someone or
      something that decides to
      transport it   39
  to happen needs someone or
      something to transport   39
  transmits disease   183
transport as natural acts   5
Transportation Communications
      International Union (TCU)   2
transport centre of the world   206
transport codes
  examination of knowledge   54
  need to know   54
  seek clarity, brevity, and an
      unambiguous message   54
transport communication theory   49
transport communications
  aesthetics   71
  dehumanize   124
  management perspective   11
  qualitative aspects   52

union perspective 11
walking talking 144
transport demand, build more
    infrastructure and vehicles 96
transport design and planning need
    multidisciplinary team 206
transport efficiency Preface
transport engineers, reassess
    standards 182
transport experience, aesthetics 71
transport growth, can be destructive
    93
transport network is two networks, fixed
    facilities and traffic 38
transport networks, all are
    communications networks 24
transport news, big stories 69
transport of hazardous things 190
transport paradigms 77
    new 96
transport paradigms and episteme of
    globalization 77
transport people think paradigmatically
    within transport modes 83
transport policy issues 13
transport security 178
transport service providers 43, 47, 158
    depends on selling a transport
        service 124
transport signage 57
    seeks optimal balance between
        information and redundancy 57
transport signage systems 52
transport signs, features of good
    signage 61
transport studies 14
    information theory implicit in 20
    lacks core body of theory 82
    paradigm 81
transport system, code of conduct 52
transport systems 11
    in search of profit 68
    manage themselves 196
    own code of signs 54
transport systems need internal
    communication systems 11
transport technology, future 96
transportation networks, fundamental
    dyad of an origin and a destination
    linked by a route 21

transportation psychology 68
Trans-Saharan Highway, Europe to
    Dakar 130
travel suggests excitement 69
travel agents 124
travel for the sake of travelling 89
travel to school 178
travelators 86
trickle-loading 187
trolley 144
trouble in transport 177, 178
tsunami 181
tsunami warning system 182
turn-around time 116
    reduction 171
two orders of meaning in
    communications 52
typhus 183

u-Japan 196
unambiguous orders 123
uncertainty, weather induced 118
uncoupling communications and
    transport 91
underground railways 131
understanding relationship between
    transport and communications 12
unexpected, allowing for the 104
unexpected information 57
UNHCR 190
unified theory of communications 6
unions 114
United Nations (UN) 9
United Nations Law of the Sea
    (UNCLOS) 121
United States 185, 195, 206
    transcontinental railways 130
uniting theory
    common basis in systems theory 12
    constructs of human logic 12
    grounds for 12
universal tracking, technically
    feasible 168
unmanned spacecraft 120
unstressed stock 172
unusual, ability to deal with the 120
urban mass transit 87
urban road transport paradigm 142
urban transport, powered by people and
    horses 87

u-scenario 195
use of ubiquitous carbon and
    hydrogen 97
user pays policies, hard for democratic
    countries to implement 135

vagrants, of no fixed abode 132
value of one bit of information 8
vehicle fleet tracking 163
vehicle inspection 98
vehicle interface skills 49
vehicle testing, motorized 136
vehicles communicate as well as
    transport 139
Venice 110
violent weather 194
virtual classes 199
virtual laboratory 24
virtual objects 101
virtual people 101
virtual pipes 205
virtual reality (VR) 75, 101
virtual schools 199
virtual space 101
virtual teleconferencing 101
virtual tools 101
virtual travel 203
virtual voices 102
virtual worlds 6
visual flight rules (VFR) 116
visual surveillance 122
vital systems monitoring 141
voice communications 165
vulnerability to EMP of
    telecommunications 180

walking is to transport what speaking is
    to communication 145
war
  speeds advances in transport
      technology 180

temporary no-go zones 180
war and genocide 179
warehouse management system
    (WMS) 165
water transport 107
wayfinding behaviour 62
wayfinding skills 141
wearable computing Preface, 98
wearable IT 137
weather 111
weather IT 182
web-browser 163
welfarism 19
wheelchairs 86
wheels as extensions of legs 156
whooping cough 183
wind drogues 114
wind energy 195
windjammers 89
wireless 163
word associations 123
work benches and tables 149
World Court 9
World Health Organization (WHO)
    184
world parliament 207
World Trade Organization (WTO) 9,
    125
World Wars 78
  impetus for transport innovation
      117
world's biggest machine 206
Wright brothers 108

yearning to be free 108
yellow Jack flag 184

zeppelin 96, 200
Zuckerman lecture 181